T0123164

Seaweeds of the British Isles

Seaweeds of the British Isles

A collaborative project of the British Phycological Society
and the Natural History Museum

Volume 1 Rhodophyta

Part 2A Cryptonemiales (*sensu stricto*)
Palmariales, Rhodymeniales

Linda M Irvine

Natural History Museum, London

First published by the Natural History Museum,
Cromwell Road, London SW7 5BD
© Natural History Museum, London, 1983

This edition printed and published by Pelagic Publishing, 2011,
in association with the Natural History Museum, London

The Authors have asserted their right to be identified as the Authors of this
work under the Copyright, Designs and Patents Act 1988.

ISBN 978-1-907807-09-1

This book is a reprint edition of 0-565-00871-4.

All rights reserved. No part of this publication may be transmitted in any
form or by any means without prior permission from the British Publisher.

A catalogue record for this book is available from the British Library.

Seaweeds of the British Isles

Volume 1 Rhodophyta

Publishing in three parts

Contents

Foreword

The second part of the *Seaweeds of the British Isles* to be published continues the critical appraisal of the marine Rhodophyta. The format and style of treatment remain as were adopted for part 1. Although it was originally intended to include the corallines in this part, the practical difficulties in dealing with this poorly understood group would have meant a long delay in the completion of the work; for this reason, it has been decided to treat them as a separate part.

Work is continuing on the other algal groups under the authorships indicated below. Every effort is being made to make these treatments available as soon as possible, but meanwhile the respective authors will much appreciate the communication of any data which may assist their work. Rhodophyta 2B (Hildenbrandiales, Corallinales) – Mrs L. M. Irvine and Dr Y. M. Chamberlain. Rhodophyta 3A (Ceramiales) and 3B (Bangiophyceae) – Prof. P. S. Dixon. Vol. 2 Chlorophyta – Dr E. M. Burrows, and Vaucheriales – Prof. T. Christensen. Vol. 3A & B Phaeophyta – Dr R. L Fletcher and Dr G. Russell. Vol. 4 Cyanophyta – Dr B Whitton.

Thanks are due to all those who have assisted in any way with the present part, by sending information and specimens, or by collaboration in the field or laboratory. It is a special pleasure to acknowledge the care and cooperation of the artists, whose illustrations form such an important feature of this Flora.

J. F. M. Cannon

Keeper of Botany
British Museum (Natural History)

ix

Preface

The first part of Volume 1 (Rhodophyta), published in 1977, was prepared jointly by Prof. P. S. Dixon and Mrs L. M. Irvine, the former having overall responsibility for the Introduction and Nemaliales and the latter that for the Gigartinales. Accounts of the remaining orders are being prepared separately by the same authors. Because of particular difficulties with the Corallinaceae and Hildenbrandiaceae (recently transferred to the Corallinales and Hildenbrandiales, respectively) the present part now excludes these families and a separate account is in preparation. Part 3, dealing with the Ceramiales and Bangiophyceae, is being written by Prof. Dixon; a supplement, key, glossary and index prepared by both authors will be included in this final part. In the meantime, a provisional glossary will be found at the end of this part.

ERRATA
Volume 1 Rhodophyta Part 1 Introduction, Nemaliales, Gigartinales
page (x)   Ulster Museum, Belfast: for BFT read BEL.
page 194 *Fucus lanceolatus* Withering: for 1976 read 1796.
page 202 Fig. 72: for D read E; for E read D.
page 252 Petrocelis: for 246 read 243.

Acknowledgements

I would like to record my gratitude to the following collaborators: Dr W. F. Farnham (Halymeniaceae), Dr M. D. Guiry (Palmariales, Rhodymeniales), Miss C. A. Maggs (Peyssonneliaceae) and Dr M. T. Martin (parasitic genera) and also to Prof. P. S. Dixon who has been a continuing source of inspiration and encouragement.

In addition to the colleagues acknowledged in Part 1, I would like to thank Mr J. F. Castle, Dr Y. M. Chamberlain (Mrs Butler), Dr C. J. Chapman, Dr I. Haugen, Mrs S. Hiscock, Mr S. I. Honey, Mr R. M. Howlett, Prof. E. B. G. Jones, Dr A. L. Rice, for the loan or gift of specimens, records etc. and Mr R. Ross and Dr P. C. Silva for help with nomenclature.

Additional to the list in Part 1, I am also indebted to the directors and curators of the following institutions for providing facilities and assistance in many ways. The herbarium abbreviations follow Holmgren *et al.* (1981); those marked * are provisional.

Société des Sciences Naturelles, Cherbourg (France)   CHE
Department of Botany, University College, Cork   CRK*
Department of Botany, University College, Galway   GALW
Scottish Marine Biological Association, Dunstaffnage, Oban   SMBA*
Department of Biological Sciences, Portsmouth Polytechnic
Station Biologique, Roscoff (France)

Most of the illustrations were prepared by Mrs V. C. Gordon Friis. The remaining plates were completed by Mrs M. C. Tebbs.

ARRANGEMENT OF THE WORK

The notes below are additional to those given in the introduction and taxonomic treatment in Volume 1 Rhodophyta Part 1, to which the reader is referred for general background information; it is cited here as 1(1). Because the geographical area under investigation is limited, it has been our policy to use arrangements of genera within higher taxa accepted by authorities who have either a more comprehensive mandate or monographic approach to the subject (e.g. Kylin, 1956). Authors of floras, check-lists, etc. do not profess to be authorities on the circumscription of higher taxa, especially those poorly represented in the area, and readers are advised to consult other works for this information.

As stated earlier (1(1) p.69), we intended this volume to follow the most recent check-list of British marine algae (Parke & Dixon, 1976) but inevitably it has been found necessary to incorporate some changes. These result both from recent publications and from our own investigations. For convenience, the changes are listed here as well as in the appropriate places in the text. Recently, Pueschel & Cole (1982, 1982a) have suggested a regrouping of the Rhodophyta into orders on the basis of an ultrastructural survey of pit plugs. They give strong evidence in support of retaining the Palmariales (see Guiry & Irvine *in* Guiry, 1978a; Christensen, 1980), reinstating the Gelidiales and Bonnemaisoniales (see 1(1) and Dixon, 1973) and for removing the Corallinaceae and Hildenbrandiaceae from the Cryptonemiales to independent orders. These last two families will be dealt with in Volume 1 Part 2B. (See Silva & Johansen, 1983; Pueschel, 1982).

The family names used are those given in Parke & Dixon (1976) with the references and proposals for conservation following Silva (1980). Particular attention has been paid to studying the little known and ill-defined noncalcareous encrusting algae formerly placed in the Squamariaceae, an obviously artificial grouping of unrelated genera (see Denizot, 1968). In co-operation with a number of colleagues, attempts have been made to elucidate their life histories and to place them in other more natural families to which they appear to be related taxonomically. They are usually sublittoral, fertile in winter and occur mixed together on a variety of substrates. They were often first noticed on pottery, glass and old shells, probably because they are more easily seen and collected on such substrates, but their occurrence is in fact more widespread. They have been distinguished on the basis of relatively few characters pertaining to the texture of the thallus, the form of the vegetative filaments and the relative position, insertion and nature of division of the tetrasporangia. Some appear to be involved in the life histories of other species (e.g. *Haematocelis rubens,* q.v.); the discovery of further such associations is to be expected world wide.

1

The following changes involving genera and species should be noted:

Callophyllis flabellata Crouan frat. to *C. laciniata* (Huds.) Kütz.

Chylocladia squarrosa (Kütz.) Le Jol. to *C. verticillata* (Lightf.) Bliding.

Cruoriopsis hauckii Batt. to *Rhododiscus pulcherrimus* Crouan frat.

Dumontia incrassata (O. F. Müll.) Lamour. = *D. contorta* (S. G. Gmel.) Rupr.

Halosacciocolax to Palmariales; *H lundii* Edelstein = *H. kjellmanii* Lund

Kallymenia larterae Holm. to *K. reniformis* (Turn.) J. Ag.

Meredithia microphylla (J. Ag.) J. Ag. = *Kallymenia microphylla* J. Ag.

Peyssonnelia rupestris Crouan frat. to *Rhodophysema elegans* (Crouan frat. ex J. Ag.) Dixon

Rhododiscus Crouan frat. to Gloiosiphoniaceae

Rhodophysema Batt. to Palmariales

Rhodymenia pseudopalmata (Lamour.) Silva var. *ellisiae* (Duby) Guiry & Hollenberg = *R. holmesii* Ardiss.

Dermocorynus montagnei Crouan frat. and *Peyssonnelia immersa* Maggs & L. Irvine included here have been recorded for the British Isles since the publication of the check-list. Any further additions and alterations will be incorporated in the supplement at the end of the volume.

The selected synonymy given for each species is usually restricted to the basionym and any other names used in the check-lists of British marine algae and Newton (1931). In cases where there is a nomenclatural or taxonomic problem it has been necessary to include a fuller synonymy.

The species descriptions are based as far as possible on personally collected material, amplified by studies of a wide range of specimens in herbaria and elsewhere. Comments and references in the text indicate where other sources have been used. For features of a general nature readers should refer to the descriptions given under genera and families since, to avoid unnecessary repetition, these are not given under each species.

Although spore or sporangium size does not appear to provide a diagnostic criterion for separating related species (see Ngan & Price, 1979), it can be very useful for distinguishing between unrelated species of similar morphology, as in foliose taxa.

Few discharged spores have been available, and so their measurements are not routinely given. The tetrasporangia have been found to exhibit a wide range in length and this can be indicated conventionally by roughly grouping them into the following categories: very large (>200 μm), large (100–200 μm), medium (50–100 μm), small (25–50 μm) and very small (<25 μm). The cystocarp diameter can be described similarly: very large (>2 mm), large (1–2 mm), medium (500 μm–1 mm), small (250–500 μm) and very small (<250 μm).

It has not always been possible to take as many measurements as we would have liked, and so some are given in more detail than others. For example, in the case of *Grateloupia doryphora*, which has been the subject of detailed investigations, the tetrasporangial sizes are given as (29)35–45(53)×(8)12–20(25) μm. In other cases, such as *Callocolax neglectus* where few carposporangia have been measured, the size is quoted simply as *c.* 10 μm.

In certain foliose genera of the Cryptonemiales (*Cryptonemia, Halymenia, Kallymenia* and to some extent *Grateloupia*) stellate cells (sometimes called ganglionic or arachnoid cells) occur at the interface between the cortex and medulla; for convenience, they are considered here to be part of the medulla. Their development is discussed by Codomier (1974). They are also known to occur in the exotic genus *Sebdenia* (Gigartinales).

Certain commonly-used terms have been avoided because of possible ambiguity, etc. For example, 'nemathecium' has been used previously in different, specialized senses: in *Polyides* (Rao, 1956), in *Peyssonnelia* (Denizot, 1968), and in *Rhodymenia* (Dawson, 1941), and so.it seems preferable to use the general term 'sorus', qualified as necessary, for any aggregation of reproductive bodies. Similarly, 'paraphysis' appears to have been used indiscriminately by phycologists. The terms 'di-' and 'polystromatic' are not used because they apply to situations found only in plants with true parenchyma.

During the compilation of distribution limits within the British Isles, constant reference has been made to Dixon, Irvine & Price (1966), Price (1967), Price & Tittley (1970, 1975) and Guiry (1978b) but, as noted in 1(1) p. 70, only verified records have been included.

The distribution of each species is given on the basis of the county system in use before 1974 with a few subdivisions as necessary. Younger readers will probably need to refer to the map included in the end papers! This system gives rise to a few anomalies in geographical limits: records in north Devon can be further north than some in Somerset, for example, and there are even greater problems in Scotland. It is hoped that these provisional, county based data will be superseded by the British Phycological Society's Mapping Scheme.

Records are based as far as possible on attached, rather than drift, specimens, although, surprisingly, this difference is not always clearcut. For example, *Grateloupia* spp. sometimes grow on small, loose stones which can be moved some distance by tidal currents and the same applies to a number of encrusting algae on unstable substrates. It is also known that loose-lying populations of certain species such as *Phyllophora crispa* occur in the shallow sublittoral in sheltered areas (Burrows, 1958), as do the beds of unattached, branched, coralline algae called maerl.

The conditions promoting growth and reproduction are increasingly being investigated both in the field and in the laboratory. Some information obtained from *in vitro* studies has been included; it has not been possible to list all the details of growth conditions, but source references have been given when available.

Comparatively little is known concerning reproductive periodicity. Fruiting records have been based largely on random records of plants found bearing reproductive bodies during different months of the year. Obvious tetrasporangia or cystocarps do not necessarily indicate the presence of ripe spores, however: they may be immature, or even already shed. In some species such as *Callithamnion hookeri* spores can be produced in a few days whilst in others such as *Furcellaria lumbricalis* they apparently take nine months to develop. In some species fertile areas are easy to see because the reproductive bodies are aggregated into sori, often with some modification of associated cells. Scattered sporangia, or more particularly spermatangia, are inevitably much less frequently recorded. More sophisticated studies have not been attempted here and the details given can be considered as departure points for future work (cf. Kain, 1982).

An attempt has been made to describe variations in thallus form and to correlate these with, for example, different habitat regimes. Unfortunately, there is little experimental evidence of the kind obtained by Dixon (1966) for *Gelidium* and *Grateloupia* spp. available to substantiate the tentative conclusions drawn from the present limited observations.

As with the Gigartinales (see 1(1)), the drawings have been standarized as far as possible. Each plate is designed to include an illustration of the habit (usually natural size), a low magnification drawing, showing external features and cystocarps where possible, and a higher magnification drawing showing internal structure and tetrasporangia. The tetrasporangia, usually as seen in transverse section, have been drawn at a standard magnification to facilitate direct comparison and give an idea of the remarkable size range encountered. This magnification is similar to that provided by the 'low power' objective ($\times 10$) of a student's microscope. Occasionally, it has been necessary to include drawings at a higher magnification as well.

REFERENCES

Burrows, E. M. 1958. Sublittoral algal populations in Port Erin Bay, Isle of Man. *J. mar. biol. Ass. U.K.* **37**: 687–703.

Christensen, T. 1980. *Algae. A taxonomic survey.* Fasc. 1. Odense.

Codomier, L. 1974. Recherches sur les *Kallymenia* (Cryptonémiales, Kallymeniacées) II. Développement des spores et morphogenèse. *Vie et Milieu* **24**, (Sér.A,): 369–388.

Dawson, E. Y. 1941. A review of the genus *Rhodymenia*, with descriptions of new species. *Allan Hancock Pacif. Exped.* **3**: 123–180.

Denizot, M. 1968. *Les algues floridées encroûtantes.* Paris.

Dixon, P. S. 1966. On the form of the thallus in the Florideophyceae, pp. 45–63 *in* Cutter, E. (Ed.) *Trends in plant morphogenesis.* London.

Dixon, P. S. 1973. *Biology of the Rhodophyta.* Edinburgh.

Dixon, P. S., Irvine, D. E. G., & Price, J. H. 1966. The distribution of benthic marine algae. A bibliography for the British Isles. *Br. phycol. Bull.* **3**: 87–142.

Guiry, M. D. 1978a. The importance of sporangia in the classification of the Florideophyceae, pp. 111–144 *in* Irvine, D. E. G. & Price, J. H. (Eds) *Modern approaches to the taxonomy of red and brown algae,* Systematics Association Special Volume **10**. London.

Guiry, M. D. 1978b. *A consensus and bibliography of Irish seaweeds.* Vaduz.

Holmgren, P. K., Keuken, W. & Schofield, E. K. 1981. *Index Herbariorum.* I, *The Herbaria of the world.* ed. 7. Regnum Vegetabile vol. 106. Utrecht.

Kain, J.M 1982. The reproductive phenology of nine species of Rhodophyta in the subtidal region of the Isle of Man. *Br. phycol. J.* **17**: 321–331.

Kylin, H. 1956. *Die Gattungen der Rhodophyceen.* Lund.

Ngan, Y. & Price, I. R. 1979. Systematic significance of spore size in the Florideophyceae (Rhodophyta). *Br. phycol. J.* **14**: 285–303.

NEWTON, L. 1931. *A handbook of the British seaweeds*. London.

PARKE, M. & DIXON, P. S. 1976. Check-list of British marine algae – third revision. *J. mar. biol. Ass. U.K.* **56**: 527–594.

PRICE, J. H. 1967. The distribution of benthic marine algae. A bibliography for the British Isles. Supplement 1. *Br. phycol. J.* **3**: 305–315.

PRICE, J. H. & TITTLEY I. 1970. The distribution of benthic marine algae. A bibliography for the British Isles. Supplement 2. *Br. phycol. J.* **5**: 103–112.

PRICE, J. H. & TITTLEY, I. 1975. The distribution of benthic marine algae. A bibliography for the British Isles. Supplement 3. *Br. phycol. J.* **10**: 299–307.

PUESCHEL, C. M. 1982. Ultrastructural observations of tetrasporangia and conceptacles in *Hildenbrandia* (Rhodophyta: Hildenbrandiales). *Br. phycol. J.* **17**: 333–341.

PUESCHEL, C. M. & COLE, K. M. 1982. Ultrastructure of pit plugs: a new character for the taxonomy of red algae. *Br. phycol. J.* **17**: 238.

PUESCHEL, C. M. & COLE, K. M. 1982a. Rhodophycean pit plugs: an ultrastructural survey with taxonomic implications. *Amer. J. Bot.* **69**: 703–720.

RAO, C. S. P. 1956. Life history and reproduction of *Polyides caprinus* (Gunn.) Papenf. *Ann. Bot. N.S.* **20**: 219–230.

SILVA, P. C. 1980. *Names of classes and families of living algae* Regnum Vegetabile vol. 103. Utrecht & The Hague.

SILVA, P. C. & JOHANSEN, H. W. 1983. Reappraisal of the order Corallinales. *Br. phycol. J.* **18**: in press.

Cryptonemiales

CRYPTONEMIALES Schmitz, *sensu stricto*

CRYPTONEMIALES Schmitz in Engler (1892), p. 21.

Thalli either crustose, or frondose and then erect or prostrate, pseudo-parenchymatous, degree of aggregation of constituent filaments ranging from loose to compact; of uniaxial or multiaxial construction.

Carpogonium formed from apical cell of special accessory filament; carposporophyte developing after transfer of zygote nucleus to an auxiliary cell also borne on an accessory filament, either close to (procarpic) or at a distance from that bearing carpogonium; each carposporangium liberating one carpospore; gametangial plant and tetrasporangial plant of similar or totally dissimilar organisation.

An order not sharply separated from the Gigartinales; with representatives of the following families occurring in marine situations in the British Isles:
Dumontiaceae
Halymeniaceae
Gloiosiphoniaceae
Kallymeniaceae
Choreocolacaceae
Peyssonneliaceae.

The family Choreocolacaceae was created by Sturch (1926) and includes several genera of supposed parasites. Carposporophyte development is not fully understood but appears to differ in different genera. Kylin (1956) considered the family to be related to the Kallymeniaceae but in some cases the auxiliary cell is said to be cut off after fertilization as in the Ceramiales. The family is here retained in the Cryptonemiales pending further investigations (see Norris, 1957).

The family Peyssonneliaceae was redefined by Denizot (1968) to include only those encrusting genera for which both gametangia and tetrasporangia were known. The genus *Haematocelis* has, however, been provisionally assigned to this family here, pending further investigations into its relationship with *Schizymenia*, 1(1) p. 175 (see André, 1977, 1980) and *Sphaerococcus*, 1(1) p. 204 (see Maggs & Guiry, 1982a). Two other genera provisionally placed in this family by Dixon & Irvine *in* Parke & Dixon (1976) have been removed, *Rhododiscus* to the Gloiosiphoniaceae and *Rhodophysema* to the Palmariaceae (Palmariales).

Recent work by Pueschel & Cole (1982, 1982a) and others suggests that the Hildenbrandiaceae and Corallinaceae should be given ordinal status. These will be dealt with in Volume 1 Part 2B (See Pueschel, 1982; Silva & Johansen, 1983).

DUMONTIACEAE Bory

DUMONTIACEAE Bory (1828), p. 197 [as Dumontiae].

Thallus with erect fronds, terete, compressed or flat; branching lateral, sometimes dichotomous or proliferous;? uniaxial, medulla filamentous, cortex compact or loose; auxiliary cells

9

remote from carpogonia, connecting filaments fusing first with cells of carpogonial branch, then with auxiliary cells, tetrasporangia zonate or cruciate.

This family is represented in the British Isles by three genera: *Dilsea, Dudresnaya* and *Dumontia*. According to Wilce (pers. comm.) the frond primordia are multiaxial, but each axial filament separates in turn to form a lateral branch. Kylin (1956) interpreted the apical development of *Dilsea* as uniaxial, but with a 'marginal meristem', saying that in very young plants the primary frond has a transversely dividing apical cell. Bert (1965) maintained that the subsequent multiaxial construction of *Dilsea* was sufficient grounds for segregating it into a separate family, Dilseaceae. Abbott (1968), did not share this view; she did, however, propose the family Weeksiaceae for those genera of the Dumontiaceae in which the gonimoblast develops from a cell of the carpogonial branch, the auxiliary cells remaining functionless. This group is not represented in the British Isles.

DILSEA Stackhouse

DILSEA Stackhouse (1809), p. 71.

Type species: *D. edulis* Stackhouse (1809), p. 71 (= *D. carnosa* (Schmidel) O. Kuntze (1893), p. 404).

Sarcophyllis Kützing (1843), p. 401.

Erect fronds arising from small, discoid base, foliose, stipitate, undivided or irregularly split; mature structure multiaxial, medulla compact, filamentous, becoming interspersed with rhizoids, cortex broad, compact with a fairly well-defined layer of larger cells inwards and radial rows of small cells outwards.

Gametangial plants probably dioecious; spermatangia in superficial sori, rarely recorded; carpogonial and auxiliary cell branches in separate branch systems, numerous, lying between cortex and medulla, many celled, branched and with associated sterile filaments, coiled; connecting filaments fusing with cell of carpogonial branch and then with auxiliary cell, gonimoblast developing inwards, most cells becoming carposporangia, enveloping filaments present; cystocarps very small, immersed, without a pore; tetrasporangia intercalary in inner cortex, cruciate.

One species in the British Isles:

Dilsea carnosa (Schmidel) O. Kuntze (1893), p. 404.

Lectotype: BM (see Turner, 1809, pl. 114) France (Dieppe).

Fucus carnosus Schmidel (1794), p. 76.
Fucus edulis Stackhouse (1801), p. 57, nom. illeg. non *F. edulis* S. G. Gmelin (1768), p. 113.
Dilsea edulis Stackhouse (1809), p. 71.

Erect fronds arising from a small, expanded base, terete stipe usually short, to 10 mm; expanding gradually into a spoonshaped blade; blade fleshy, opaque, up to 500 mm long and usually about half as broad in the widest part; colour reddish brown, bleaching to brick-red; younger blades entire, older blades split, sometimes with eroded margins; usually about 600–700 μm thick, up to 900 μm in fertile regions.

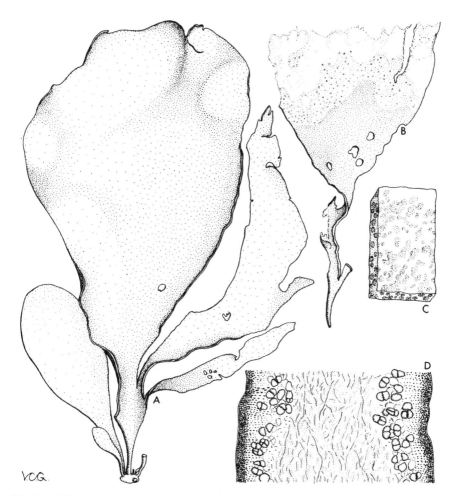

Fig. 1   *Dilsea carnosa*
A. Habit (Feb.) × ⅔; B. Habit of plant releasing carpospores (Feb.) × ⅔; C. Part of same showing embedded cystocarps (Feb.) × 8; D. T.S. blade with tetrasporangia (Feb.) × 80.

Mature structure multiaxial; medulla compact, composed of branched longitudinal thick-walled filaments about 10 μm in diameter, becoming interspersed with rhizoids; cortex consisting inwards of rounded cells about 20 μm in diameter and outwards of closely packed radial filaments of about 6 cells, gradually diminishing in size to about 5 μm, polygonal in surface view.

Gametangial plants dioecious; spermatangia in large, pale patches near margins of frond in younger parts, cut off from superficial mother cells, spermatia 1–2 μm, produced consecutively, remaining embedded in a layer of mucilage up to 10 μm thick and appearing in rows; gonimoblasts developing inwards in medulla, sometimes lobed or coalescing, cystocarps very small, up to 250 μm, grouped near the margins in younger parts, giving the frond the texture of sandpaper, enveloping filaments and pore absent; carposporangia about 45×30 μm, spores released by thallus decay; tetrasporangia in obscure sori near the margins in younger parts, reputedly intercalary in the cortical filaments, 45–65×30–50 μm, spores cruciately arranged.

Epilithic; upper sublittoral to at least 24m.
Generally distributed throughout the British Isles.
Spitsbergen; Arctic Russia to Portugal (Rio Douro); ?Greenland.

Perennial; spermatangia recorded for June, carpogonial branches recorded from May to December, cystocarps maturing in March; tetrasporangia recorded from January–April.
This is one of the species least variable in external appearance.
Levring *et al.* (1969) report that this species is eaten in some coastal regions.
For a description of a pyrenomycete parasite, see Maire & Chemin (1922).

D. edulis has a similar internal structure to *Schizymenia dubyi* (Chauv. ex Duby) J. Ag. but is thicker, see 1(1), p. 178. Young plants can be confused with *Kallymenia microphylla* J. Ag. q.v.; the latter can be distinguished by the well developed stellate cells in the medulla.

DUDRESNAYA Crouan frat. nom. cons.

DUDRESNAYA Crouan frat. (1835), p. 98.

Type species: *D. coccinea* (C. Agardh) Crouan frat. (1835), p. 98 (= *D. verticillata* (Withering) Le Jolis (1863), p. 117).

Borrichius S. F. Gray (1821), p. 317, 330.

Erect fronds arising from small discoid base, terete, much branched, very lubricous; structure uniaxial, apical cell transversely divided, axial filament visible above, becoming invested by a mass of rhizoids below, and producing whorls of 4 branched radial filaments loosely aggregated in mucilage.

Gametangial plants monoecious; spermatangia terminal on radial filaments; carpogonial branches curved, 7–9-celled, auxiliary cell branches 12-celled with central auxiliary cell, connecting filament fusing first with 2 cells of carpogonial branch, then with auxiliary cell, gonimoblast developing outwards, cystocarps immersed, very small, almost all cells becoming carposporangia, enveloping filaments and pore absent; tetrasporangia scattered on radial filaments, zonate.

One species in the British Isles:

Dudresnaya verticillata (Withering) Le Jolis (1863), p. 117.

Lectotype: BM-K. An unlocalized, undated Velley specimen accepted provisionally to be of this status.

Ulva verticillata Withering (1796), p. 127.
Ulva coccinea Poiret in Lamarck & Poiret (1808), p. 165, nom. illeg. non *U. coccinea* Hudson (1778), p. 567.
Dudresnaya coccinea (C. Agardh) Crouan frat. (1835), p. 98.

Erect fronds arising from small discoid base, terete, fragile, very lubricous, translucent, up to 200 mm long and increasing in diameter with age to 2 mm, colour rose-pink to dark red or brownish, axis or a few main branches percurrent, outline broadly pyramidal, bushy, much branched, branching irregularly alternate, patent, appearing banded when young.

Structure uniaxial, apical cell dividing transversely, axial filament bearing radiating whorls of 4 delicate, repeatedly dichotomous filaments composed inwards of somewhat clavate cells about 10 μm in diameter, decreasing to about 3 μm in the outermost cells which may be up to 25 μm long, axial filament visible only in the youngest parts, soon becoming obscured by branched intertwined rhizoids produced by the inner cells of the radial filaments as the thallus matures.

Gametangial plants monoecious; spermatangia colourless, borne terminally on the radial filaments, 2–3 μm in diameter; cystocarps without enveloping filaments, developing among inner radial filaments, very small, up to 125 μm in diameter, scattered throughout the plant,

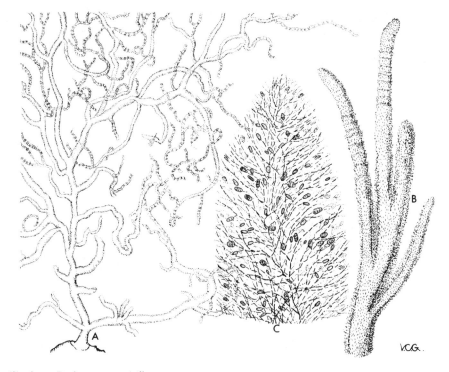

Fig. 2   *Dudresnaya verticillata*
A. Habit (Aug.) × 1; B. Branch apex (Aug.) × 8; C. Branch apex with tetrasporangia (Aug.) × 80.

pore absent, carposporangia 18–20 μm, released by separation of the radial filaments; tetrasporangia produced laterally on the radial filaments, 25–53×11–23 μm, with spores zonately arranged.

Epilithic and occasionally epiphytic; upper sublittoral to at least 13m, often on pebbles, maerl and gravel, tolerant of tidal streams.

Widely distributed on southern and western shores, northwards to Orkney and eastwards to Kent; Aberdeen.

Norway (Oslo) and W. Sweden southwards to France (Brest); Mediterranean; Canary Isles.

Annual, growth beginning in early spring though plants not usually conspicuous until June; becoming fertile in July and disappearing in October; spermatangia recorded for June–August, cystocarps and tetrasporangia for July–August, sometimes occurring on the same individual.

There is some variation in size of plants and degree of branching but this has not been investigated.

The anatomical differences between *D. verticillata* and other species with a similar lubricous habit are discussed under *Calosiphonia vermicularis* (J. Ag.) Schmitz (see 1(1), p. 171). It has been suggested (S. Hiscock pers. comm.) that *D. verticillata* is an opportunist summer annual growing in places where pebbles are stable in summer but moved about in winter preventing establishment of perennials. There is a characteristic assemblage of species in such places including *Radicilingua thysanorhizans* (Holmes) Papenf., *Scinaia forcellata* Biv., *S. turgida* Chemin and *Stenogramme interrupta* (C. Ag.) Mont.

DUMONTIA Lamouroux

Dumontia Lamouroux (1813), p. 133 (reprint p. 45).

Type species: *D. incrassata* (O. F. Müller) Lamouroux (1813), p. 133 (reprint p. 45). (= *D. contorta* (S. G. Gmelin) Ruprecht (1850), p. 298.)

Thallus consisting of a discoid holdfast and terete erect fronds later becoming compressed and hollow, unbranched or irregularly laterally branched; mature structure uniaxial, apical division usually transverse, axial filament visible only at apex, medulla of large cells interspersed with down-growing filaments, cortex compact with inner cells larger than outer.

Gametangial plants dioecious; spermatangia produced from outer cortical cells in large superficial sori; carpogonial branches and auxiliary cell branches numerous in inner cortex, carpogonial branches 4–7-celled, auxiliary cells central in 4–6-celled branches, connecting filaments fusing with several auxiliary cells, gonimoblast developing outwards, almost all cells becoming carposporangia, enveloping filaments and pore absent, cystocarps immersed; tetrasporangia lateral on inner cortical filaments, cruciate.

One species in the British Isles:

Dumontia contorta (S. G. Gmelin) Ruprecht (1850), p. 298.

Lectotype: original illustration (Gmelin, 1768, pl. 22 fig. 1, see Abbott, 1979) in the absence of specimens. Sea of Okhotsk.

Fucus contortus S. G. Gmelin (1768), p. 181.
Ulva incrassata O. F. Müller (1775), p. 7, non *U. incrassata* Hudson (1778), p. 572, nom. illeg.
Dumontia incrassata (O. F. Müller) Lamouroux (1813), p. 133 (reprint p. 45).

Thallus consisting of a discoid holdfast up to 25 mm in diameter from which erect fronds arise endogenously, usually in groups; fronds fairly soft, becoming somewhat lubricous, narrow and terete at first, becoming hollow, inflated and twisted when old, dark brownish red, often bleaching to bright or pale yellow, up to 230 mm long and 15 mm broad, irregularly laterally branched below, not or rarely repeatedly so; branches markedly attenuate at point of insertion, apices tapering but not acuminate.

Structure uniaxial when mature; medullary filaments variable in diameter, 10–20 μm, often thick-walled, medulla distended, hollow and filled with mucilage, cortex of short radial filaments, outermost cells irregular about 6–12×6 μm in surface view.

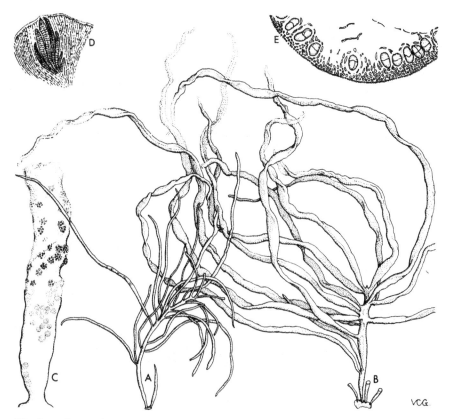

Fig. 3   *Dumontia contorta*
A. Habit of young plant (May) × 1; B. Habit of older plant (Aug.) × 1; C. Branch with mature cystocarps (June) × 8; D. part of T.S. with tetrasporangia (May) × 80; E. Part of basal crust showing endogenous origin of erect frond (Oct.) × 80.

Gametangial plants dioecious, spermatangia *c.* 3 μm in diameter, cut off in pairs from outer cortical cells, in irregular sori sometimes covering entire plant; gonimoblast developing outwards from a large fusion cell, all cells becoming carposporangia about 45–60×30–40 μm, cystocarps very small, about 150 μm, containing 20–30 sporangia, enveloping filaments and pore absent; tetrasporangia scattered in inner cortex 55–90 × 35–65 μm with a thick (10 μm) wall, spores cruciately arranged.

Epilithic, rarely epiphytic, in pools and on rock, upper littoral to upper sublittoral in sheltered and exposed areas, tolerant of silt, reduced salinity and insolation.

Generally distributed throughout the British Isles.

Arctic Russia to Portugal (Douro); Baltic; Arctic Canada to USA (New York); NW Pacific, Alaska. Probably circumpolar in the Arctic (see Hoek, Van den, 1982).

Erect fronds annual, endogenous, primordia visible within the crust as early as August; fronds elongating and becoming recognisable on the shore from November onwards, becoming fertile in April/May and dying back from apex as spores are shed; spermatangia recorded for April, cystocarps from May–July and tetrasporangia from May–August. Knight & Parke (1931) reported a difference in the size and behaviour of plants at different levels on the shore. In northeastern USA similar differences were observed by Whelden (1928), and Setchell (1923) found that a water temperature of 10–20°C was necessary to initiate fruiting, Kilar & Mathieson (1978) found that, in New Hampshire, plants grew rapidly in the coldest months and reached their maximum stature before becoming fertile. Spermatangia occurred from March–April, carpogonia from March–May, cystocarps from April–June and tetrasporangia from March–August. For comments on the behaviour of plants in Denmark see Kristiansen (1972); for culture observations see Rietema (1982); Rietema & Breeman (1982); Rietema & Klein (1981). Encrusting bases have been observed (as numerous bright red spots 1–3 mm in diameter) in the field in the British Isles from May onwards. Whelden reported that if the erect fronds appear in autumn the bases do not then increase in size; in other cases the bases persist and may become more extensive and the erect fronds appear later, presumably when conditions become suitable.

Erect fronds often becoming inflated, distorted and twisted when old, especially in sheltered areas; such plants have been distinguished as *D. filiformis* var. *crispata* (Grev.) Grev. An unusual form having very elongated fronds with groups of branches near the frond apices has been found occasionally in the sublittoral (Dorset and Channel Isles).

Brebner (1895) reported that an erect frond originates by intercalary division of filaments within the basal crust, but reexamination of his slides (BM 1195, 1196) suggests that the divisions are lateral rather than intercalary.

The brown alga *Ulonema rhizophorum* Foslie is a common epiphyte; the endophytic marine fungus *Olpidium laguncula* H. E. Petersen was recorded from the Isle of Man by Lodge (see Aleem, 1955); Baardseth & Taasen (1973) described a new species of diatom, *Navicula dumontiae* living exclusively in the mucilage in the centre of the medulla.

This species is used as food in Japan and for funoran production in E. Asia (Levring *et al.*, 1969).

Young plants can be confused with *Cystoclonium purpureum* (Huds.) Batt. (see 1(1), p. 196).

HALYMENIACEAE

by
Linda M. Irvine
and
William F. Farnham*

HALYMENIACEAE nom. cons. prop.

HALYMENIACEAE Bory (1828), p. 158 [as Halymeniae], nom. illeg. (but see Silva, 1980).
Cryptonemiaceae (J. Agardh) Decaisne (1842), p. 359 [as Cryptonemeae].
Grateloupiaceae Schmitz in Engler (1892), p. 21.

Thallus with erect fronds, terete, compressed or flattened, branching variable; multiaxial, medulla filamentous; carpogonial branches and auxiliary cells in separate specialized branch clusters called ampullae, developing secondarily in cortex; gonimoblasts fairly small, produced in large numbers scattered throughout or restricted to special reproductive portions, most cells becoming carposporangia; pericarp not or only slightly elevated, with a pore; tetrasporangia in sometimes thickened outer cortex, scattered or in sori, cruciate.

Three genera are known to occur in the British Isles, *Cryptonemia*, *Dermocorynus* and *Grateloupia*. The genus *Dermocorynus* has only recently been recorded for the British Isles; it is similar to *Grateloupia* in both anatomy and reproduction but is distinguished by the very small size of the erect frond (see Kraft, 1977, Guiry & Maggs, 1982a). A fourth genus, *Halymenia*, is represented by *H. latifolia* Crouan frat. in northern France and a few sterile specimens that may belong to this species have been found in the British Isles (but see Maggs & Guiry, 1982b). For comments on generic criteria in the family, see Kraft (1977) who considers that features of habit and vegetative anatomy are likely to be more important taxonomically than differences in the development of the carposporophyte.

CRYPTONEMIA J. Agardh

CRYPTONEMIA J. Agardh (1842), p. 100.

Type species: *C. lactuca* sensu J. Agardh (1842), p. 100 (= *C. lomation* (Bertoloni) J. Agardh (1851), p. 227).

Thallus erect, stipitate below, flattened above into a thin, entire, dichotomously branched or palmately lobed blade often with an evanescent midrib; structure multiaxial, fairly compact and strong, medullary filaments interwoven with a network of rhizoids, highly refractive stellate cells present, cortex compact with larger cells inwards and smaller cells outwards, not arranged in distinct rows.

* Marine Laboratory, Portsmouth Polytechnic, Ferry Road, Hayling Island, Hants. PO11 0DG

Spermatangia unknown; carpogonial and auxiliary cell branches separated, each in a special compact branch system (ampulla), carpogonial branches 2-celled, supporting cell intercalary in primary filament, gonimoblasts developing outwards from a small, basal fusion cell, most cells becoming carposporangia, with a few enveloping filaments, cystocarps small, either scattered over surface or confined to secondary bladelets which are usually small and terminal, in local thickenings of inner cortex, protruding both into medulla and externally; tetrasporangia scattered in cortex, cruciate, sometimes occurring only in secondary bladelets.

KEY TO SPECIES

Stipe comparatively long, to 45 mm, black, often winged, primary and secondary
　　blades triangular, somewhat lobed and rounded distally .... *C. seminervis*
Stipe very short, to 5 mm, red, not winged, primary blade strap-shaped, secondary
　　blades subdichotomous, ovate or obovate, often with a distinct apical point　*C. hibernica*

Cryptonemia hibernica Guiry & L. Irvine (1974), p. 225.

Holotype: BM. Paratypes: DBN. Cork (Camden).

Thallus with a discoid holdfast about 10 mm in diameter and erect fronds to 600 mm long, cerise to brownish red, stipe terete or somewhat compressed, up to 5 mm long and 1–2 mm in diameter, red, expanding into a blade with a wedge-shaped thickened portion below; blades sometimes narrow and entire but usually to 100 mm broad and 450 mm long or more, ovate or obovate, often with a cordate base, pointed at the apex, subdichotomously lobed, with an undulate, slightly thickened margin; further blades produced as proliferations from the margins and occasionally from the surface in older plants; blade thickness 80–165 μm, increasing to 200–250 μm fertile, and 350 μm in thickened, areas.

Structure multiaxial; medullary filaments 4–6 μm in diameter, interspersed with a network of thicker, irregular, highly refractive cells lying in the plane of flattening of the blade; cortex of radial filaments of usually 3 or 4 cells, to 10 cells in thickened areas.

Spermatangia unknown; gonimoblast developing outwards, most cells becoming carposporangia, 10–14 μm, with a few enveloping filaments, cystocarps very small, scattered, elevating the cortex, 100–300 μm in diameter with an obscure pore; tetrasporangia scattered in the cortex, regularly cruciately divided, c. 14×12 μm.

Epilithic on bedrock, pebbles and shells, rarely on *Laminaria* holdfasts; sublittoral to at least 12m in areas of silty sand, strong current and turbid water.

Recorded only for Ireland (County Cork: Kinsale, Oysterhaven and Cork Harbour).

Perennial, growth occurring throughout the year with a continuous production of new blades and periodic erosion followed by proliferation from remnants. Spermatangia unknown, cystocarps recorded for August and from November–February, tetrasporangia recorded for August and November.

Blades show much variation in appearance from linear-lanceolate to broad, lobed and ruffled. Cullinane & Whelan (1981) considered that blade-shape depends more on habitat than seasonal variation as suggested by Guiry & Irvine (1974), the linear-lanceolate plants being found on mobile substrates such as shells and pebbles. The bullate appearance described by Guiry & Irvine seems to be a feature of drift material not shown by attached plants.

Fig. 4   *Cryptonemia hibernica*
A. Habit of plants on shell (July) × ⅔; B. Portion of blade with cystocarps (July) × 17;
C. T.S. blade with tetrasporangia (July) × 80; D. Surface of blade showing tetrasporangium and stained medullary filaments (July) × 300; E. Squash preparation of stellate
medullary cells × 100.

C. hibernica has affinities with a group of northeastern Pacific species. The possibility of its being an introduced species was discussed by Guiry & Irvine. It can be distinguished from other foliose species by the thickening at the base of the blade, typical of the genus.

Cryptonemia seminervis (C. Agardh) J. Agardh (1876), p. 165.

Lectotype: LD (Herb. Alg. Agardh. 22959; see Agardh, 1822a pl. XVII). Spain (Cadiz).

Sphaerococcus seminervis C. Agardh (1822a), p. 2.

Erect fronds to 100 mm long, arising from a discoid holdfast about 6 mm in diameter; stipes terete, black, up to 45 mm long and about 1 mm in diameter, with blade-like wings, expanding above into a narrow plane fan-shaped blade into which the stipe extends as an evanescent midrib, rose-red, up to 80 μm thick with thickened, slightly undulate margins and irregular in outline above.

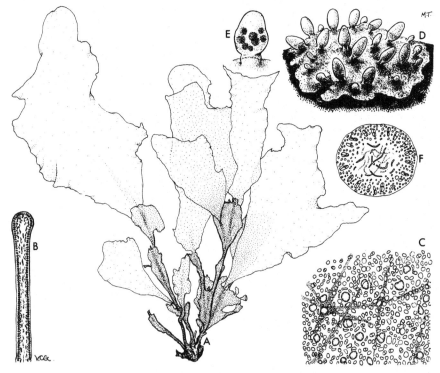

Fig. 5   *Cryptonemia seminervis*
A. Habit (Sep.) × 1; B. T.S. part of blade with tetrasporangia (Sep.) × 80; C. Surface view of blade with tetrasporangia and stellate medullary cells (Sep.) × 300.
Dermocorynus montagnei
D. Habit of plant showing erect fronds (papillae) arising from crust on rock (Oct.) × 4; E. Erect frond with cystocarps (Oct.) × 16; F. T.S. erect frond with tetrasporangia (July) × 80.

Structure multiaxial; medullary filaments c. 3 μm in diameter, stellate cells very highly refractive and conspicuous with arms 3–5 μm in diameter, cortex of 1–2 layers of very small, rounded cells 3–6 μm in surface view.

Gametangial plants unknown; tetrasporangia scattered in cortex, 10–12×8–10 μm, spores cruciately arranged.

No data on habitat for the British Isles; reported by Ardré (1970) to occur in Portugal on exposed sandy rocks in shady places in the upper sublittoral and in shady lower littoral pools. Channel Isles (unattached) to southern Spain; Morocco.

No data on seasonal behaviour or form variation in the British Isles is available.

The record for the British Isles is based on a drift specimen collected in Herm, Channel Isles by den Hartog on 7 September 1960, (see Dixon, 1961, as *C. lactuca* (C.Ag.) J. Ag.). *C. seminervis* can be disguished from other species with a similar habit by the thickening (evanescent mid-rib) at the base of the blade and the stellate cells in the medulla.

DERMOCORYNUS Crouan frat.

DERMOCORYNUS Crouan frat. (1858), p. 69.

Type species: *D. montagnei* Crouan frat. (1858), p. 70.

Thallus with a comparatively thick firm crust and small erect papillae bearing reproductive organs; crust composed of a basal layer of cells and sparingly branched erect filaments, rhizoids absent; papillae multiaxial, medulla distinctly filamentous, cortex compact, of small-celled branched filaments.

Gametangial plants monoecious; spermatangia in superficial sori formed singly or in pairs on a mother cell derived from a cortical cell; carpogonial branches and auxiliary cells developing in separate ampullary systems, carpogonial branches 2-celled, carpogonium and hypogynous cell giving rise to connecting filaments which fuse with auxiliary cells, gonimoblast developing outwards, forming one or two lobes surrounded by a thin layer of enveloping filaments; cystocarps very small, with a pore; tetrasporangia terminal in cortex, scattered over whole surface of papillae, cruciate.

One species in the British Isles:

Dermocorynus montagnei Crouan frat. (1858), p. 70.

Lectotype: CO. France (Rade de Brest).

Crusts to 30 mm or more in extent, to 220 μm in thickness, brownish red, reproductive papillae few to many, erect, brownish red, unbranched, to 2 mm long and 500 μm broad.

Crust consisting of a strongly adherent basal layer of radially expanded, occasionally branched cells each of which gives rise to an erect filament which only branches once at the second or third cell from the base, rhizoids absent, cells of erect filaments strongly coherent, secondary pit connections absent; cells with a distinctly polygonal outline in surface view, 5–10 μm in diameter, in vertical section 5–7 μm long, covered with mucilage to 8 μm thick; papillae multiaxial, medulla distinctly filamentous, surrounded by 1–2 layers of stellate cells 15–25(40)×10–20 μm, cortex of 4–6 layers of regularly dichotomously branched cells decreasing in size outwards, secondary pit connections frequent, surface cells polygonal, 5–8 μm in diameter.

Gametangial plants monoecious; spermatangia in sori, colourless, 4–8 × 2–4 μm; carposporangia 6·5–12 μm in diameter, mature cystocarps very small, 75–100 μm in diameter, formed at middle and base of papillae, immersed, with an indistinct small pore, two gonimoblast lobes usually present surrounded by a few indistinct enveloping filaments; tetrasporangial sorus formed over most of the surface of a papilla, raised, mucilaginous, tetrasporangia (25)35–38×10–18 μm, with cruciately arranged tetraspores.

Epilithic on small stones, sublittoral, 3–9 m.
Galway, Clare.
British Isles and France (Brittany).

Cystocarpic and tetrasporangial plants were found in Galway Bay from May to September (Guiry & Maggs, 1982a). Reproductive papillae were produced continuously after about 30 weeks in culture at 16°C under a 16:8 h photoregime in morphologically similar gametangial and tetrasporangial phases. Culture studies have also shown that the species can withstand relatively high light intensity.

Papillae containing tetrasporangia are regular in outline whilst those containing cystocarps are more irregular and warty when mature.

This species has been found recently in the British Isles (Guiry & Maggs, 1982a). It appears to be confined to small stones in the sublittoral where its small size makes it very inconspicuous and it resembles juvenile erect fronds of several species. It has an anatomy like that of *Grateloupia* but, unlike small plants of *G. dichotoma* J. Ag. or *G. filicina* (Lamour.) C. Ag., the frond is always terete.

GRATELOUPIA C. Agardh nom. cons. prop.

Grateloupia C. Agardh (1822), p. 221.

Type species: *G. filicina* (Lamouroux) C. Agardh (1822), p. 223, (but see Parkinson, 1981, 1981a, 1981b).

Thallus with erect fronds, compressed or flattened, dichotomously branched, often proliferous, mostly in plane of flattening, rarely foliose and undivided; structure multiaxial, medulla a network of loose filaments interwoven with narrower rhizoids and surrounded by larger cells some of which have elongated arms lying mainly in plane of flattening of blade (stellate cells); cortex compact, of small cells in short radial rows.

Gametangial plants monoecious or dioecious, spermatangia scattered or in small sori, developing from the outer cortical cells; carpogonial and auxiliary cell branches separated, each in a special compact branch system (ampulla), carpogonial branches 2-celled, auxiliary cell near base of its branch, gonimoblasts developing outwards with a few enveloping filaments, cystocarps small, scattered, embedded or slightly protruding, with a pore; tetrasporangia in cortex, usually scattered, sometimes in sori, cruciate.

Grateloupia is a genus in which the species are notoriously difficult to define, making the geographical ranges of each difficult to assess. The stellate cells which occur in the outer medulla in this genus are similar to, but not as elaborate as, those found in the genera *Cryptonemia, Halymenia* and *Kallymenia* where they have highly refractive proteinaceous contents.

G. filicina var. *luxurians* and *G. doryphora* appear to be introductions into British waters (see Farnham, 1980). In culture, plants of *G. filicina* var. *filicina* and *G. filicina* var. *luxurians* have each retained their own morphologies suggesting that there is some genetic difference between them. We have refrained from considering them as distinct species, however, because of the existence of plants intermediate in appearance in the Channel Isles and, more commonly, in the Mediterranean.

KEY TO SPECIES

1 Thallus foliose, more than 20 mm broad, up to 1 m or more long .... *G. doryphora*
  Thallus not foliose, not more than 10 mm broad even when over 500 mm long 2
2 Thallus up to 750 mm long, soft .... *G. filicina* var. *luxurians*
  Thallus up to 80(120) mm long, tough 3
3 Branching appearing ± pinnate or if dichotomous then main axes not more than
  1 mm broad, terete to compressed, greenish to blackish purple, sometimes red............
  .... *G. filicina* var. *filicina*
  Branching dichotomous, never appearing pinnate, axes at least in part 1·5 mm or
  more broad, flattened, reddish brown, never purple .... *G. dichotoma*

Grateloupia dichotoma J. Agardh (1842), p. 103.

Lectotype: LD (Herb. Alg. Agardh. 22637). France (near Nice).

Erect fronds arising from a discoid base up to 5 mm in diameter, stipe short, to 1 mm, expanding slightly into a repeatedly dichotomous flattened strapshaped blade, lower divisions often broader above than below but apical ones tapering, terminal dichotomies usually divergent; frond mucilaginous but firm, bright to dark reddish brown, up to 50(80) mm long, 2–3(5) mm broad, up to 230 μm thick in fertile regions.

  Structure multiaxial, medulla rather loose, composed of branched elongated filaments about 4–5 μm in diameter, becoming interspersed with narrower rhizoids, surrounded by larger cells rounded in transverse section, sometimes with elongated arms, cortex a layer 4–8 cells thick, the outermost closely packed, angular and 4–6 μm in diameter in surface view.

  Gametangial plants dioecious; spermatangia not found in British material; elsewhere, in scattered colourless sori, cut off from the outermost cortical cells, about 2 μm in diameter; cystocarps immersed in upper branches, very small, 180×220 μm with a few evanescent to persistent enveloping filaments, carpospores 14–18 μm, released through a pore 30–40 μm in diameter; tetrasporangia scattered in the cortex, 28–35×15–25 μm, spores cruciately arranged.

  Epilithic on vertical walls of deep pools in upper sublittoral, extending to at least 9 m on cobbles in gullies, often with *G. filicina* var. *filicina*.

  Recorded from southwestern shores, eastwards to Dorset, northwards to N. Devon and Somerset; Channel Isles.

  British Isles southwards to Canary Isles; Mediterranean, Black Sea; Caribbean. There are two main centres of distribution: Mediterranean and West Indies, with scattered records for Ceylon, Samoa and Japan. The species is probably more widespread than present records suggest.

  Usually annual, occasionally overwintering, resulting in larger plants the following summer. Reproducing mainly in summer, cystocarps recorded for July–September, and tetrasporangia for March, July–August; spermatangial plants have been found in Brittany in November.

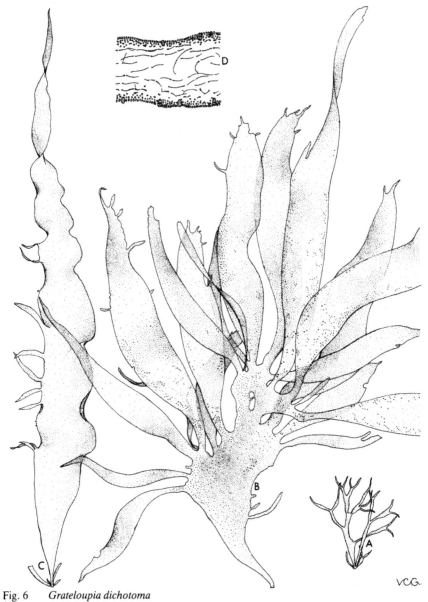

Fig. 6   *Grateloupia dichotoma*
A. Habit (July) × 1
Grateloupia doryphora
B., C. Habit of two plants (Aug.) × 1; D. T.S. blade with tetrasporangia (June) × 80.
(Cystocarps of these species as in Fig. 7).

In the British Isles plants have been found with up to 5 dichotomies. There is otherwise comparatively little variation in external appearance and the species is comparatively easy to recognise.

De Valéra's report for Kerry (see Dixon 1966) is based on small specimens of *G. filicina* but it is likely that *G. dichotoma* will be found in Ireland. Plants found in the deeper sublittoral are usually less than 10 mm long but frequently are fertile at this size; sterile plants can be confused with *Chondrus crispus* Stackh., but the latter has a conspicuous thick external mucilaginous layer (to 5 μm). Herbarium specimens are occasionally misidentified as *Scinaia* spp. (see 1(1) p. 145) or *Gymnogongrus* spp. (see 1(1) p. 217). Plants of *Scinaia* show a distinctive pattern of large and small cortical cells in surface view, however, whilst *Gymnogongrus* has a large-celled, pseudoparenchymatous medulla.

Grateloupia doryphora (Montagne) Howe (1914), p. 169.

Holotype: PC (Herb. Montagne). Peru (Callao).

Halymenia? doryphora Montagne (1839), p. 21.
Halymenia lanceola J. Agardh (1841), p. 19.
Grateloupia lanceola (J. Agardh) J. Agardh (1851), p. 182.

Erect fronds arising in groups of 1–6 from a discoid base up to 15 mm in diameter, with a usually simple stipe to 25 mm long and 2–4 mm in diameter, expanding into a large linear to broadly lanceolate, often asymmetrical blade which is sometimes longitudinally split and frequently bears blade-like proliferations from margins and surface; mucilaginous but firm, slippery; brownish red to deep crimson; up to 1 m or more long and 20–200 mm broad, usually about 450–650 μm thick.

Structure multiaxial, medulla rather loose, filaments 4–6 μm in diameter, running in all directions, interspersed with narrower rhizoids and surrounded by stellate cells 7–20 μm in diameter with elongated arms to 100 μm long; cortex 3–10 cells thick, outermost cells radially elongated, 3–8 μm in surface view.

Gametangial plants monoecious, spermatangia isolated, cut off singly from the outermost cortical cells, 3–4 μm in diameter; cystocarps small, to 300 μm, most cells becoming carposporangia 13–15 μm, enveloping filaments evanescent, pore prominent, to 100 μm when spores shed; tetrasporangia scattered in outer cortex, (29)35–45(53)×(8)12–20(25) μm, spores cruciately arranged.

Epilithic on small (to 60 mm) loosely embedded stones, upper sublittoral, rarely to 7 m; usually in sheltered areas in shallow pools and harbour drainage channels but occasionally withstanding some wave action; tolerant of periodic salinity variation from 10–36‰ and highly eutrophic conditions.

Known only from a few localities in mainland Hampshire and Sussex.

British Isles; Portugal to Ghana; Angola. Apparently widely distributed and recorded elsewhere under a variety of names.

Perennial; cystocarps and tetrasporangia recorded throughout the year, with a peak in the summer with over 90 per cent of individuals then fertile; sporelings can be found throughout the year but are most abundant in summer, and these overwinter producing mature plants the following year.

In areas sheltered from wave action the blades remain entire with few proliferations; elsewhere they become variously subdivided and older plants may show regeneration after damage.

Although it is one of the largest red algae in the British Isles, this species was not collected until 1969 (see Farnham & Irvine, 1973). It may have been present for much longer in secluded parts of Chichester and Pagham Harbours where the species flourishes in high water temperatures (to 25°C). It occurs only in mainland Hampshire and W. Sussex and does not appear to be extending its range in the British Isles; it is surprisingly absent from the Isle of Wight. Foliose plants belonging to the genus *Grateloupia* have been recorded for most warmer seas; they are probably all conspecific with *G. doryphora*, see Dawson, Acleto & Foldvik (1964) and Ardré & Gayral (1961). Gayral (1958) commented that in Morocco the species (as *G. lanceola*) was especially luxuriant when growing near sites of organic pollution.

Spermatangia have been found only rarely in the British Isles; elsewhere (e.g. Venezuela) they have been reported as occurring in superficial sori. British plants of *G. doryphora* were at first identified tentatively as a species of *Kallymenia* or *Schizymenia*. In *Schizymenia* the cystocarps contain fewer larger carposporangia not surrounded by enveloping filaments. *G. doryphora* differs from other foliose algae such as species of *Kallymenia* and *Halymenia* in lacking the well-developed highly refractive stellate cells which occur in the medulla in those genera.

This species is commercially used as carrageenan and funoran raw material in the Pacific (Levring *et al.*, 1969).

Grateloupia filicina (Lamouroux) C. Agardh (1822), p. 223.

G. filicina var. *filicina*

Lectotype: original illustration (Wulfen, 1789, pl. 15, fig. 2) in the absence of material (see Dixon, 1959). Adriatic.

Fucus filicinus Wulfen (1789), p. 157 nom. illeg. non *F. filicinus* Hudson (1762), p. 473.
Delesseria filicina Lamouroux (1813), p. 125 (reprint p. 37).
Gelidium corneum var. θ *flexuosum* Harvey (1846), pl. LIII.
Grateloupia minima Crouan frat. (1859), p. 142.

Erect fronds arising from a discoid base up to 20 mm in diameter, simple or dichotomous when young, frequently proliferous from the margins when older and appearing pinnate, mucilaginous but firm, translucent, reddish, greenish or blackish purple, often shiny when dry; up to 40(120) mm long, 2–5 mm broad, terete to compressed, usually 150–200 μm thick, proliferations tapering at both ends.

Structure multiaxial, medulla rather loose, filaments about 7 μm in diameter, mainly longitudinal, interspersed with narrower rhizoids surrounded by larger cells with some development of stellate arms; cortex 3–8 cells thick, outermost 3–7 μm in diameter in surface view.

Gametangial plants monoecious, spermatangia not found in the field in the British Isles, elsewhere and in culture scattered, developing from outer cortical cells, 4–5 μm in diameter; cystocarps very small, up to 210 μm, often confined to the proliferations, most cells becoming carposporangia 14–23 μm, enveloping filaments few, evanescent, pore 40–60 μm in diameter; tetrasporangia scattered in cortex, sometimes confined to frond apices (26)30–40(47)×(11)15–20(27) μm, with cruciately arranged spores.

Epilithic on bedrock and large stones, littoral in pools and sublittoral to 10 m, with some plants in higher pools in late summer and autumn; tolerant of some sand cover and a wide range of exposure to wave action.

Widely distributed on southern shores, northwards to Anglesey, eastwards to Norfolk; in Ireland northwards to Donegal and eastwards to Waterford; Channel Isles.

British Isles to S. Africa; Mediterranean; USA (N. Carolina) to Brazil. Cosmopolitan in all warmer seas except Australasia.

Usually annual; young plants (to 5 mm long) are commonly found from January–May, increasing to 40 mm or more in August with 1–3 dichotomies and often marginal proliferations, frequently overwintering to February/March; individuals can become fertile at any size, sometimes when only 5 mm long; such plants are often encountered in the sublittoral. Spermatangia not recorded for the British Isles; cystocarps recorded throughout the year, most abundant in the summer; tetrasporangia recorded throughout the year. In culture at 20°C carpospores produced erect fronds within 3 weeks and tetrasporangia in 6 weeks. Tetrasporelings produced cystocarps in 14 weeks, the carpospores being shed after 24 weeks.

Fronds may be unbranched (= *G. minima*), dichotomous (and then sometimes misidentified as *G. dichotoma*) or apparently pinnate (see Dixon, 1966): these variants may be found even in the same gathering. Plants occasionally bear short facial proliferations, a feature more common in var. *luxurians*. Large plants which are intermediate between the two varieties have been found in the Channel Isles (e.g. Dixon, 1961, to 120 mm long).

Lucas (1950) said that the previous report of *G. filicina* for the Netherlands was based on a specimen of *Cystoclonium purpureum* (Huds.) Batt.; his own report of a specimen on drift *Himanthalia* is also doubtful.

Perithecia of the ascomycete *Chadefaudia gymnogongri* (J. Feldm.) Kohlm. have been found in plants from England, Ireland, France and Spain. It has not been found outside Europe or on var. *luxurians* (see Farnham & Jones, 1973).

Very small plants could be mistaken for *Dermocorynus montagnei* Crouan frat., q.v.

Grateloupia filicina var. ***luxurians*** A. & E. S. Gepp (1906), p. 259.

Lectotype: BM. Australia (Sydney).

Erect fronds arising from a discoid base up to 20 mm in diameter, compressed or flattened, simple or dichotomous when young, proliferations arising from margins, blades appearing once or twice pinnate when older, facial proliferations often also present; mucilaginous but firm, brownish red to deep crimson, greenish in high insolation, up to 700 mm long, 10 mm broad and usually about 1–3 mm thick, proliferations compressed, tapering at both ends.

Structure multiaxial, medullary filaments 5–7 μm, interspersed with narrower filaments 2–4 μm running in all directions, stellate cells not as well developed as in *G. doryphora*; cortex of 10–14 cells thick, outermost radially elongated, 7–14 μm, 4–5 μm in surface view, with an external mucilaginous layer up to 40 μm thick.

Gametangial plants monoecious; spermatangia in superficial sori 50–60 μm in diameter; mature cystocarps over 300 μm with a pore increasing to 90 μm, carpospores 11–20 μm when shed, tetrasporangia (29)30–40(60)×(8)13–16(18) μm, with spores cruciately arranged.

Epilithic, in lower littoral lagoons, harbours, estuaries in sheltered running water, sublittoral to 6 m; tolerant of periodic salinity variation from 12 to 37‰.

Dorset, Hampshire (including Isle of Wight) and Sussex.

Mediterranean; Gulf of Guinea to S. Africa; Caribbean; widely distributed in the Indo-Pacific.

Attachment disc perennial, fronds overwintering, usually losing many branches but able to regenerate; growth rate 20–30 mm per month in spring, increasing to 50–150 mm per month in June–September. Spermatangia recorded for January, carposporangia and tetrasporangia

Fig. 7   *Grateloupia filicina* var. *filicina*
A. Habit (Aug.) × 1
Grateloupia filicina var. *luxurians*
B. Habit (Aug.) × 1; C. Part of branch with cystocarps (Aug.) × 8. (Anatomy and tetrasporangia of this species as in Fig. 6)

recorded throughout the year with a peak in June–August, but seasonal variability not well defined; cystocarps can persist for 2–3 months before discharging spores.

Fronds very variable in width and in number and size of proliferations; sometimes spirally twisted, often showing regeneration after damage.

This variety was first recorded for the British Isles by Farnham & Irvine (1968). The earliest specimen traced was collected in 1947 by A. H. Norkett from Bembridge, Isle of Wight, as *Helminthocladia*; other specimens were found misidentified as *Calliblepharis, Dumontia* or *Halarachnion*. Despite its wide range of form, it is readily distinguished from these by its branching pattern and small cruciate tetrasporangia. Farnham & Irvine considered it to be an immigrant; it does not appear to be extending its range in the British Isles and has not been recorded for the Channel coast of France (Farnham, 1980). It may be more widespread in the Atlantic than present records suggest: for a discussion of possible synonyms see Børgesen (1935) and Chapman & Parkinson (1974).

Used as food and as a carrageenan raw material in the Pacific (see Levring *et al.*, 1969).

HALYMENIA C. Agardh nom. cons. prop.

HALYMENIA C. Agardh (1817), p. XIX.

Type species: *H. floresia* (Clemente) C. Agardh (1822) p. 209.

Isymenia J. Agardh (1899), p. 60.
Hymenophloea J. Agardh (1899), p. 67.

Thallus with erect fronds terete, compressed or flat, variously branched and sometimes also proliferous, rarely unbranched, blade mucilaginous; structure multiaxial, medulla a loose network of filaments mostly at right angles to plane of flattening and forming bridges from cortex to cortex, surrounded by cells with stellate arms lying in plane of flattening; cortex narrow, inner cells larger than outer.

Gametangial plants where known monoecious; spermatangia in small superficial sori; carpogonial branches 2-celled, remote from auxiliary cell branch systems, gonimoblast developing inwards, not lobed, most cells becoming carposporangia, enveloping filaments present, cystocarps small, scattered, protruding into medulla and externally, with a pore; tetrasporangia scattered in cortex, cruciate.

Halymenia latifolia Crouan frat. ex Kütz. (see Dixon & Irvine, 1970) has been reported for scattered localities in the British Isles, e.g. Antrim (Batters, 1900; Newton, 1931 fig. 167) and the Isle of Man (Kain, 1960). These two reports are in fact based on broad specimens of *Halarachnion ligulatum* (Woodw.) Kütz., see 1(1) p. 184. Other records, for which specimens have not been traced, are probably based on similar misidentifications. Recently, Maggs & Guiry (1982b) have found sublittoral plants of *Halymenia latifolia* in Galway Bay and have given a full description of its morphology and phenology.

A number of collections of sublittoral foliose plants with stellate cells in the medulla remain unidentified at present because they are either sterile or bear only tetrasporangia. They have been found in various parts of the British Isles from Cornwall to Shetland and appear to represent several different species. They may belong to the genus *Halymenia* but stellate cells also occur in *Kallymenia*, q.v., and in *Sebdenia* (Gigartinales), a genus not so far recorded in the British Isles. (See Abbott, 1967, Blunden *et al.*, 1981 and Codomier, 1972, 1974).

GLOIOSIPHONIACEAE Schmitz

GLOIOSIPHONIACEAE Schmitz in Engler (1892) p. 21.

Thallus with or without erect fronds, erect fronds terete with whorled branching or compressed with lateral branching; uniaxial, medulla distinctly filamentous, cortex fairly compact; carpogonial and auxiliary cell branches borne on same supporting cell (procarpic), connecting filament fusing directly with auxiliary cell, gonimoblasts developing outwards, surrounded only by displaced cortical filaments, with or without a pore, most cells becoming carposporangia, tetrasporangia among erect filaments of an encrusting plant or occasionally reputedly in cortex of erect fronds, cruciate.

The type genus, *Gloiosiphonia*, occurs in the British Isles. It is known to have an encrusting phase in which tetrasporangia are borne (Edelstein, 1970; Edelstein & McLachlan, 1971). Several genera have been described on the basis of crusts which resemble that of *Gloiosiphonia* and one of these with a representative in the British Isles, *Rhododiscus*, is included here because it seems to be closely related to *Gloiosiphonia* (but see Maggs, Guiry & Irvine, 1983). Another, *Plagiospora*, has been placed here provisionally because of its similar tetrasporangia, although its affinities may prove to lie elsewhere.

GLOIOSIPHONIA Carmichael

GLOIOSIPHONIA Carmichael in Berkeley (1833), p. 45.

Type species: *G. capillaris* (Hudson) Carmichael in Berkeley (1833), p. 45.

Gametangial thallus erect, arising from a small disc, terete, much branched, mucilaginous but firm, older plants hollow; structure uniaxial, axial filament bearing whorls of 4 radial, branched filaments forming a compact outer cortex peripherally, down-growing rhizoids investing axis to form medulla and loose inner cortex.

Monoecious or dioecious; spermatangia produced superficially from outer cortical cells; carpogonial branches developing from basal cell of branch system, 3-celled, auxiliary cell branches forming part of same branch systems, auxiliary cells intercalary, connecting filaments not necessarily fusing with auxiliary cell of same branch system; gonimoblast developing outwards, most cells becoming carposporangia, enveloping filaments and pore absent.

Thallus of tetrasporangial phase crustose and closely similar to plants described as *Rhododiscus pulcherrimus* Crouan frat., tetrasporangia produced terminally on erect filaments, cruciate.

The gametangial phase of species of *Gloiosiphonia* produces comparatively large, much-branched, erect fertile fronds in summer. In culture carpospores from such plants have been found to produce small, exclusively crustose, plants in which tetrasporangia are borne (but see Morohoshi & Masuda, 1980; DeCew, West & Ganesan, 1981).

One species in the British Isles:

Gloiosiphonia capillaris (Hudson) Carmichael in Berkeley (1833), p. 45.

Lectotype; original description (Hudson, 1778, see Irvine & Dixon, 1982) in the absence of material. Kent (Sheerness), Devon and Cornwall.

Fucus capillaris Hudson (1778), p. 591.

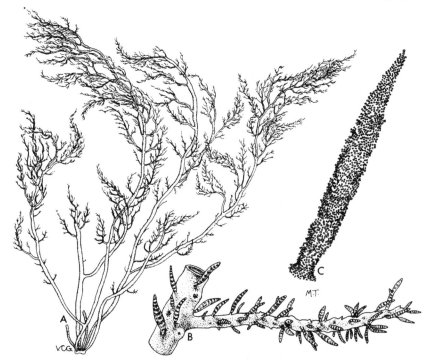

Fig. 8   *Gloiosiphonia capillaris*
A. Habit (April) × 1⅓; B. Branch of mature plant with cystocarps (July) × 8; C. Branchlet showing uniaxial apex (Aug.) × 80.

Gametangial fronds arising in groups from an expanded encrusting base up to 3 mm in diameter, mucilaginous but firm, pale to deep purplish red, old fronds sometimes hollow, to 350 mm in length, terete, to 5 mm in diameter, bushy and often pyramidal in outline, axis percurrent, repeatedly branched, branching alternate, branches attenuate at both extremities; developing from a uniaxial filament of large elongated cells which produce whorls of 4 radial branched filaments loosely packed with down-growing rhizoids, cell size decreasing outwards, radially elongated, 8–13 μm in surface view in mature fronds.

Monoecious; spermatangia terminal on the cortical filaments, grouped together to form minute whitish sori; cystocarps lying among the cortical filaments, c. 700 μm, enveloping filaments absent, carpospores released by pushing apart of cortex, c. 16 μm; tetrasporangia occurring in separate encrusting plants and, reputedly, rarely in erect fronds, terminal on short erect filaments, obliquely or irregularly cruciately divided, 20–24×12–20 μm.

Plants with erect fronds epilithic in shallow pools in lower littoral and sublittoral to 5 m, in conditions of moderate exposure to wave action.

Plants with erect fronds occurring sporadically throughout the British Isles; not recorded for southeast England between Hampshire and Lincolnshire, apart from the original record from Sheerness (Hudson, 1778) which Tittley & Price (1977) were not able to confirm.

Plants with erect fronds recorded from Iceland, Norway (Trondelag) to northern Portugal; Mediterranean. Canada (Newfoundland) to USA (Connecticut). North Pacific?

For comparisons and distribution of crustose phases, see Edelstein (1970), Edelstein & McLachlan (1971), Hollenberg (1971) and under *Rhododiscus pulcherrimus* Crouan frat.

Erect phase short-lived, recorded only between April and September; occurrence spasmodic, in some localities abundant at irregular intervals. Spermatangia recorded for July, cystocarps for May–August; encrusting phases probably perennial.

There is comparatively little variation in the appearance of erect fronds apart from length, which can be up to 350 mm.

Edelstein (1970) compared her tetrasporangial crusts with Newton's illustration (1931, fig. 181) of *Cruoriopsis hauckii* Batt., which may be a synonym of *Rhododiscus pulcherrimus* Crouan frat., q.v. Similar crusts have been found in California by Hollenberg (1971). Although there are several reports of tetrasporangia in erect fronds (e.g. Ekman, 1857, Areschoug, 1875, Kylin, 1907, Goor van, 1923), none of these has been confirmed. In some cases they appear to be based on tetrasporangia in epiphytic species such as *Ceramium* (Landsborough, 1844, Lyngbye, 1819, as *Gigartina lubrica*); others are possibly endophytes (cf. *Audouinella brebneri* (Batt.) Dixon 1(1) p. 83) or the products of *in situ* germination of carpospores. Culture studies in Japan (Morohoshi & Masuda, 1980) have shown that the crusts grown from carpospores can produce erect gametangial plants directly; see also DeCew, West & Ganesan (1981).

Gametangial plants have sometimes been confused with *Lomentaria clavellosa* (Turn.) Gaill. In the former, the branchlets are tapered at both ends, whereas in the latter the apices are more rounded. Except in very young plants, the small, crowded, naked cystocarps of *G. capillaris* can be seen as bulges in the cortex, those of *L. clavellosa* being borne externally and having a well-developed, conspicuous pore. Frankland, finding the two species on the rocks at Scarborough, commented that the former has 'the tint of lake' whilst the latter was yellow-red (see Turner, 1808).

The structure of *G. capillaris* is also similar to that of *Calosiphonia vermicularis* (J. Ag.) Schmitz, see 1(1), p. 170 and *Naccaria wiggii* (Turn.) Endl., see 1(1), p. 150.

PLAGIOSPORA Kuckuck

PLAGIOSPORA Kuckuck (1897), p. 393.

Type species: *P. gracilis* Kuckuck (1897), p. 393.

?*Cruoriopsis* Luigi Dufour (1864), p. 59.

Thallus encrusting, closely adherent to substrate, basal layer 1–2 cells thick, giving rise to erect filaments embedded in soft mucilage.

Gametangia unknown; tetrasporangia lateral on erect filaments, obliquely cruciate.

This genus may be synonymous with *Cruoriopsis*, described earlier by Dufour (1864). A character frequently considered to be of taxonomic importance is the position of the tetrasporangia; they are lateral in both *P. gracilis* and *C. crucialis*, the type species of *Cruoriopsis*. An isotype slide in BM of the latter shows laterally borne but regularly cruciate tetrasporangia about 40×20 μm. The specimen is probably not identical with *Cruoriella armorica* Crouan frat., as supposed by Denizot (1968), (see also Masuda, 1976). Kuckuck, on the other hand, characterised *Plagiospora* on the oblique division of the tetrasporangia which he had seen before only in *Hildenbrandia*. This kind of division has been reported in other encrusting species such as *C. hauckii* Batt., *C. danica* Rosenv. (see discussion under *Rhododiscus pulcherrimus*) and in the tetrasporangial crust of *Gloiosiphonia capillaris* Carm. in Berk., q.v. Personal observations indicate that it would be premature to compare the taxonomic validity of these features without further studies on the structure and behaviour of these small encrusting algae and their involvement with the life histories of other species.

One species in the British Isles:

Plagiospora gracilis Kuckuck (1897), p. 393.

Lectotype: PC (material at Biologische Anstalt Helgoland destroyed during Second World War). Helgoland (Nordhafen).

Cruoriopsis gracilis (Kuckuck) Batters (1902), p. 95.

Crust dark ruby red, drying shiny and slightly rugulose, mucilaginous, up to 10 mm in diameter and 150 μm thick, rhizoids absent; prostrate filaments 1–2 layered, giving rise to little-branched erect filaments loosely united by mucilage and easily separated by gentle pressure, usually about 20 cells long, cells beadlike, 3–5 μm broad, as long as broad or sometimes longer, cell fusions present.

Gametangial plants unknown; tetrasporangia lateral on erect filaments, 12–17×6–9 μm, obliquely cruciately divided.

On stones and pottery in the sublittoral, recorded to 15 m.
S. Devon and Galway. Isle of Man report not confirmed.
Norway (Oslofjord), British Isles, Helgoland, W. Baltic; Canada (Newfoundland) and USA (Maine).

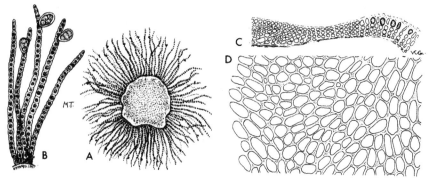

Fig. 9   *Plagiospora gracilis*
A. Tetrasporangial crust on sand grain (Dec.), squash preparation × 80; B. Part of same showing lateral tetrasporangia × 300.
Rhododiscus pulcherrimus (Type)
C. Crust with undivided terminal tetrasporangia (Jan.) × 80; D, surface of crust × 300.

Tetrasporangia recorded for November–March in the British Isles, reported at other times elsewhere.

Data on form variation insufficient for comment.

This species was first recorded by Batters (1896) and Brebner (1896) as '*Cruoriopsis cruciata* Zan.' (= *C. crucialis* (Luigi Dufour) Zan.) but the identification was corrected later (Batters, 1902).

Rosenvinge (1917) described Danish specimens as larger, to 15 mm, bright purple and with mature tetrasporangia 21–22×11–14 µm.

RHODODISCUS Crouan frat.

RHODODISCUS Crouan frat. (1859), p. 289.

Type species: *R. pulcherrimus* Crouan frat. (1859), p. 290.

Thallus encrusting, closely adherent to substrate, rhizoids absent, lobed, somewhat mucilaginous; basal layer monostromatic, consisting of branched radiating filaments giving rise to short, almost unbranched erect filaments not easily separable under pressure, cell fusions absent.

Gametangia unknown; tetrasporangia in superficial sori, terminal on erect filaments, cruciate, sterile filaments absent.

The terminal, cruciate tetrasporangia are the distinguishing feature of the genus. In the type material, they are regularly divided, as illustrated by the Crouans. Material which is similar vegetatively but has obliquely cruciate or irregularly divided tetrasporangia has been placed in the genus *Cruoriopsis* Luigi Dufour by Batters (1896) (but see also *Plagiospora* Kuck.). Originally, however, the genus *Cruoriopsis* was based on material with immersed lateral, rather than terminal, tetrasporangia.

Rhododiscus pulcherrimus Crouan frat. (1859), p. 290.

Lectotype: CO France (Brest).

?*Cruoriopsis hauckii* Batters (1896), p. 384 pro parte non *Cruoriella armorica* sensu Hauck (1885), p. 31
   nec Rosenvinge (1917), fig. 109, 110.
?*Cruoriopsis danica* Rosenvinge (1917), p. 184.

Crusts smooth, to 8 mm in extent and up to 130 µm thick when fertile, carmine red, drying paler, mucilaginous in the region of the tetrasporangia; consisting of a monostromatic basal layer of branched radial filaments of hexagonal or elongated cells 16–20 µm long and about 7–11 µm in diameter, cell fusions absent, 1 or 2 erect filaments arising from each basal cell, few-celled, occasionally branched, firmly united but separable under pressure, and not much tapering, the cells shorter than broad to slightly longer, 10–16 µm in diameter.

Gametangial plants unknown; tetrasporangial sori becoming raised when mature, tetrasporangia developing from the terminal cells of the erect filaments, 20–25×12–15 µm, with cruciately arranged spores.

On stones, old shells, pottery and maerl, sublittoral, recorded to 16 m.

S. Devon, W. Inverness, Clare, Galway.

Recorded for the British Isles and northern France, but probably more widely distributed.

Tetrasporangia recorded for January–May, September–October.

Data on form variation too inadequate for comment.

Maggs, Guiry & Irvine (1983) reported that, in culture, tetraspores of field collected plants of *R. pulcherrimus* grew into *Atractophora hypnoides* Crouan frat. (Nemaliales, Naccariaceae), see 1(1), p. 148.

The name *Cruoriopsis hauckii* was given by Batters (1896) to a plant from the Adriatic misidentified by Hauck (1885) as *Cruoriella armorica* Crouan and also to material dredged by Brebner from the Plymouth breakwater in January 1896. Batters's description includes vegetative details of Hauck's plant, the identity of which remains obscure, together with details of tetrasporangia from Brebner's material. Batters prepared a number of slides from Brebner's material (BM 7996–8001) and these indicate it was heterogeneous; slides 7997, 8000, 8001 show a thin crust with immersed terminal tetrasporangia 20–30×15–20 µm, the dimensions given by Batters, whilst the remaining slides show a thicker crust with more elongate tetrasporangia, 20–30×7–11 µm, borne in a superficial mucilaginous sorus. Unfortunately, Batters appears to have sent a slide of this latter plant as '*C. hauckii*' to Rosenvinge (cf. 1917, p. 186 fig. 109): it is possibly a species of *Peyssonnelia*. Furthermore, the plant Rosenvinge examined from Hauck's herbarium (cf. 1917, p. 187 fig. 110) was not from the Adriatic but from Naples; it is structurally different from the Adriatic plant and agrees closely with an isotype slide in BM of *Cruoriopsis crucialis* Luigi Dufour (1864). (See Schiffner, 1916, Feldmann, 1939 and discussion under *Plagiospora*). This latter is the type species of the genus *Cruoriopsis* and has laterally borne, regularly cruciate tetrasporangia about 40× 20 µm. It is probably not identical with *Cruoriella armorica* Crouan frat., as supposed by Denizot (1968); see also Masuda (1976). Børgesen (1929, p. 11) gave a new name, *Cruoriopsis rosenvingii*, to this plant examined by Rosenvinge, applying the binomial *C. hauckii* to the material collected at Plymouth; the name should be retained for the material on slides 7997, 8000, 8001 since this agrees more closely with Batters's description. This material was illustrated by Newton (1931, fig. 181) and is closely similar to *R. pulcherrimus* (cf. Denizot, 1968). Edelstein (1970) compared this illustration with her tetrasporangial crusts of *Gloiosiphonia capillaris* (Huds.) Carm. in Berk. q.v. A Pacific species, *Thuretellopsis peggiana* Kylin (Dumontiaceae), also has similar tetrasporangial crusts (see 1(1), p. 216). Rosenvinge, misled by both Batters's and Hauck's material, described his own as a new species, *Cruoriopsis danica*. In fact, *C. danica* is very similar to the true *C. hauckii* (i.e. slides 7997, 8000, 8001) and it is interesting to compare Rosenvinge's illustrations of *C. danica* (1917, fig. 107, 108) with that of a young plant of *G. capillaris* given by Kuckuck in Oltmanns (1904). (See also Printz, 1926).

KALLYMENIACEAE W. R. Taylor nom. cons. prop.

KALLYMENIACEAE W. R. Taylor (1937), pp. 248, 274.
Callymeniaceae Kylin (1928), p. 56.

Thallus with erect fronds either compressed or flattened and foliose, branched or unbranched, entire, lobed or split; uniaxial, medulla distinctly filamentous or pseudoparenchymatous and interspersed with rhizoids, cortex compact; carpogonial branches 1 or more per supporting cell in small branch fascicles developing in cortex, auxiliary cells in same (procarpic) or separate branch system (non-procarpic), gonimoblast developing outwards, most cells becoming carposporangia, cystocarps with a cortical pericarp, somewhat protruding externally, scattered or restricted to special fertile areas, pore present or absent, tetrasporangia scattered in cortex, cruciate (but see *Kallymenia microphylla*).

The genera *Callophyllis*, *Callocolax* (parasitic on *Callophyllis*) and *Kallymenia* occur in the British Isles.

CALLOCOLAX Schmitz ex Batters

CALLOCOLAX Schmitz ex Batters (1895), p. 318.

Type species: *C. neglectus* Schmitz ex Batters (1895), p. 318.

Thallus reputedly parasitic on *Callophyllis*, cushions irregularly lobed with a lower portion penetrating into host tissue; anatomy similar to that of host.

Spermatangia unknown; carpogonial branches 3-celled with one or two 1-celled sterile branches, supporting cell functioning as auxiliary cell, gonimoblast developing outwards, without enveloping filaments, cystocarps immersed, slightly protruding, without a pore; tetrasporangia scattered in cortex, cruciate.

One species in the British Isles:

Callocolax neglectus Schmitz ex Batters (1895), p. 318.

Lectotype: BM. Isotypes: Algae Britannicae Rariores Exsiccatae No. 154. Dorset (Weymouth).

Thallus reputedly parasitic on *Callophyllis laciniata* (Huds.) Kütz., occurring on the face and margin of the host, cushion-like or irregularly lobed, 2–4 mm in extent, colour pale and brownish owing to the reduction of chloroplasts.

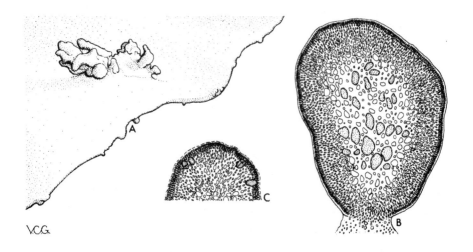

VCG.

Fig. 10   *Callocolax neglectus*
A. Habit of erect fronds on blade of *Callophyllis laciniata* (Aug.) × 8; B. V.S. frond with carposporangia (Aug.) × 80; C. V.S. frond with tetrasporangia (Sep.) × 80.

Structure multiaxial; medullary cells very thick walled, up to 75 μm in diameter, occurring together with narrow branched filaments of short pigmented cells; cortex of smaller cells *c.* 4 μm in surface view.

Spermatangia unknown; cystocarps embedded, slightly protruding, without a pore or enveloping filaments, occupying almost the entire frond, carposporangia *c.* 10 μm; tetrasporangia scattered in the cortex, 20–33×11–18 μm, spores cruciately arranged.

Sublittoral to at least 30 m; rarely on specimens of the host from the upper sublittoral. Recorded for most of the British Isles where the host occurs. Norway (Hordaland) to N. W. Spain.

Recorded for February and May–December, probably present within host plants throughout the year. Cystocarps recorded for May, June, September and October, tetrasporangia for August, September and November.

Specimens with cystocarps are dark and globular; those with tetrasporangia are lighter in colour and more lobed.

Kylin (1930) pointed out that it is difficult to confirm penetration of the parasite into the host because the medulla in the blade of *Callophyllis* spp. consists of large cells accompanied by small-celled branched filaments resembling parasitic filaments. Chemin (1937) could find no demarcation between 'parasite' and 'host' and thought they might not, in fact, be separate entities.

CALLOPHYLLIS Kützing

CALLOPHYLLIS Kützing (1843), p. 400.

Type species: *C. variegata* (Bory) Kützing (1843), p. 401.

Euthora J. Agardh. (1848), p. 11.
Nereidea Ruprecht (1850), p. (63) 255.
Crossocarpus Ruprecht (1850), p. (72) 264.
Rhodocladia Sonder (1852), p. 679.

Thallus with erect fronds, compressed, flat or foliose, branching more or less dichotomous, sometimes pinnate in narrower species; structure multiaxial, medulla compact, pseudoparenchymatous, with large cells interspersed with branched filaments composed of small cells with chloroplasts, cortex of 1–5 layers of cells becoming smaller outwards.

Gametangial plants dioecious; spermatangia in pale sori scattered over surface of blade; carpogonial branches 3-celled, with 1- or 2-celled sterile branches, apparently monocarpogonial in the British Isles, supporting cell functioning as auxiliary cell, gonimoblasts developing inwards or outwards, most cells becoming carposporangia, cystocarps immersed, protruding on one or both sides, with enveloping filaments and one or more pores, scattered over blade surface or confined to margins or marginal outgrowths, tetrasporangia immersed in cortex, scattered, cruciate.

For a taxonomic appraisal of the genera *Callophyllis* and *Euthora,* see Hooper & South (1974). *Callophyllis flabellata* Crouan frat. (1867, p. 143), originally described from France (Brest), has been reported a number of times for the British Isles, e.g. Anon (1952), Newton (1931) and Batters (1902). Specimens supporting these reports do not, however, resemble the isotype of *C. flabellata* in BM. Bert (1967) described the differences he found between *C. flabellata* and *C. laciniata* in the vicinity of Brest. The former is said to have a thinner, brittle frond with wedgeshaped divisions, crenulate apices, usually no marginal proliferations and large cystocarps (to 2 mm). It is apparently restricted to waters with much suspended matter. Specimens fitting this description have been found frequently in certain silty areas of Hampshire. Typical *C. laciniata* is absent from these areas. In these specimens the ultimate divisions of the blade are less than 100 μm thick, the cortex is 1-layered and the penultimate divisions are more elongate than those in typical *C. laciniata*. *Callocolax neglectus* Schmitz ex Batt. has not been found on these plants but its absence could be due to the silty conditions.

KEY TO SPECIES

Plants up to 30(60) mm long, main divisions rarely more than 2 mm broad; restricted to northeast British Isles ....*C. cristata*
Plants up to 150(250) mm long, main divisions usually not less than 4 mm broad; recorded throughout British Isles .... *C. laciniata*

Callophyllis cristata (C. Agardh) Kützing (1849), p. 747.

Lectotype: LINN 1274.69. Iceland.

Fucus cristatus Linnaeus ex Turner (1808), p. 48, nom. illeg., non *F. cristatus* Withering (1796), p. 103.
Sphaerococcus cristatus C. Agardh (1817), p. XVI, 29.
Euthora cristata (C. Agardh) J. Agardh (1848), p. 12.

Thallus with erect fronds arising from a small attachment disc, cartilaginous, more or less fanshaped in outline, pinkish to dark wine or brownish red, to 60 mm in length, flattened, to 2(3) mm wide below, much branched, branching irregularly alternate or dichotomous, branches narrowing markedly above giving the frond a finely dissected appearance, axils narrowly rounded.
   Structure multiaxial; medulla of large cells, sometimes with a few filaments of small pigmented cells especially near the margins and in older parts; cortex of a usually single layer of much smaller cells, closely packed, 6–9 μm in diameter in surface view.

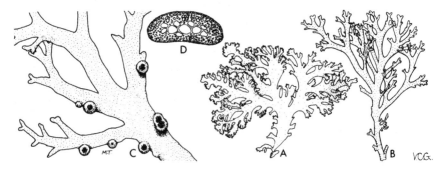

Fig. 11   *Callophyllis cristata*
A.,B. Habit of two plants (Aug.) × 1⅓; C. Branch with cystocarps (July) × 8; D. T.S. tetrasporangial plant (June) × 80.

Gametangial plants dioecious; spermatangia forming irregular superficial sori on the younger parts; cystocarps protruding, scattered along the margins of the frond, about 500 μm in diameter, with an obscure pore, gonimoblast developing outwards, carposporangia in groups surrounded by filaments, 13–20 μm; tetrasporangia crowded in the cortex in the youngest parts, 20–35×16–23 μm, spores 15–18 μm, irregularly cruciately arranged.

Epiphytic on the haptera and rarely the stipes of *Laminaria*, upper sublittoral to at least 30 m, in sheltered and moderately wave-exposed areas, in full salinity.

Restricted to the northeast of the British Isles, from the Shetland Isles southwards to at least Northumberland; reports from the Isle of Wight and the Isle of Man were based on misidentifications, other reports dubious.

In the northern Atlantic from British Isles, Denmark and USA (New Jersey) northwards; apparently circumpolar. Often recorded in the northern Pacific under the name *Euthora fruticulosa* (Rupr.) J. Ag.

Plants occurring throughout the year, probably perennial. Hooper & South (1974) report that in Newfoundland spermatangia appear only for a few weeks in the year (mid-March to mid-April); cystocarps recorded in the British Isles from June–October and tetrasporangia recorded from June–July. Dixon (1961a) reported the occurrence of tetrasporangia and cystocarps on the same thallus in plants from New Brunswick, Canada.

Considerable variation is shown in degree of dissection and width of branches. According to Hooper & South (1974), 'the most flattened, broad forms occur in severely exposed localities while the narrow, more sparsely branched forms occur in deeper and more sheltered habitats.'

This species is similar in habit to small plants of *Sphaerococcus coronopifolius* Stackh. (see 1(1), p. 205) and *Plocamium cartilagineum* (L.) Dixon (see 1(1), p. 203).

This species may occur more frequently in the sublittoral in the northeast of the British Isles than has hitherto been supposed. Much larger plants are common in colder waters of the North Atlantic. Hooper & South (pers. comm.) report the occurrence of a species of *Callocolax* on Newfoundland material of *C. cristata*.

Callophyllis laciniata (Hudson) Kützing (1843), p. 401.

Lectotype: BM. An unlocalised, undated Hudson specimen accepted as of provisional lectotype status (see Irvine & Dixon, 1982). Yorkshire.

Fucus laciniatus Hudson (1762), p. 475.

Thallus consisting of a small attachment disc giving rise to an erect stipe about 2 mm long which expands immediately into a fanshaped blade up to 150 mm (250 mm) long; blade subcartilaginous, opaque, rose to brownish or purplish red, repeatedly branched into wedgeshaped divisions 10–30 mm broad and about 150–300 μm thick; branching dichotomous to irregularly palmate, apices rounded, smooth, margins often with curled or fringed proliferations.

Structure multiaxial; medulla compact, consisting of a mixture of large colourless cells 40–150 μm in diameter and narrow, branched filaments of short pigmented cells about 5 μm in diameter; cortex consisting of about 5 layers of compact cells which become progressively smaller outwards, 2–6 μm in diameter in surface view.

Gametangial plants dioecious; spermatangial sori scattered over the blade, pale, spermatangia c. 3 μm; gonimoblasts developing towards medulla, carposporangia in clusters or

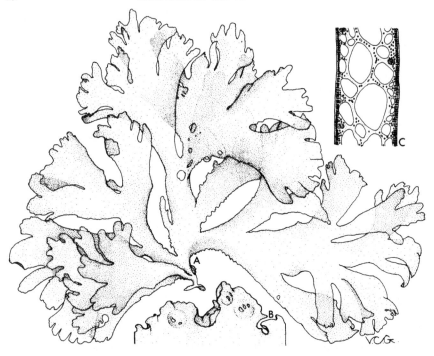

Fig. 12   *Callophyllis laciniata*
A. Habit (June) × 1; B. Part of blade with cystocarps (Nov.) × 8; C. T.S. blade with tetrasporangia (Nov.) × 80.

rows, spores 15–20 μm, enveloping filaments present, cystocarps up to 500 μm in diameter, in thickened marginal proliferations which enlarge and may be very crowded, cortex elevating on one or both sides, one or more pores present; tetrasporangia scattered, giving the surface a watermarked appearance, developing laterally from an inner cortical cell and enlarging after division to 30–40(70)×18–25 μm, spores cruciately arranged.

Epilithic and epiphytic on *Laminaria* stipes; from the upper sublittoral to at least 30 m, in both sheltered and exposed areas.

Not known for southeast England between Hampshire and Suffolk except for a single specimen from Worthing, otherwise generally distributed in the British Isles.

Faroes; Norway (Trondelag) and Sweden to Morocco; Mediterranean.

Plants recorded throughout the year, probably perennial; spermatangia recorded only for September but possibly more frequent, cystocarps mainly in the summer months with a peak in September, tetrasporangia recorded from April to November, with a peak in September.

The appearance of the frond is variable, depending on the frequency and extent of dissection and the width of the resulting divisions.

The species has occasionally been confused with *Cryptopleura ramosa* (Huds.) Kylin ex Newton, with which it often grows. It can be distinguished by the absence of iridescence and the thicker, veinless blade showing very small cells in surface view. Plants are frequently infected with the parasite *Callocolax neglectus* Schmitz ex Batt., q.v.

KALLYMENIA J. Agardh

Kallymenia J. Agardh (1842), p. 98.

Type species: *Kallymenia reniformis* (Turner) J. Agardh (1842), p. 99.

Euhymenia Kützing (1843), p. 400.
Meredithia J. Agardh (1892), p. 73.
Dactylymenia J. Agardh (1899), p. 50.

Thallus erect with an expanded, irregularly lobed or split blade, often proliferating from margin, occasionally perforated; structure multiaxial, obviously filamentous, medullary filaments loosely interspersed with rhizoids, surrounded by larger cells some of which have elongated processes, cortex compact with smaller cells outwards. Spermatangia rarely recorded; carpogonial branches solitary or several on each supporting cell, 3-celled, auxiliary cells scattered in cortex, gonimoblast developing outwards, arising from connecting filament, with a compound lobed fusion cell, cystocarps scattered over all or part of blade, elevating cortex on one or both sides, enveloping filaments present, pore absent, most cells becoming carposporangia; tetrasporangia scattered in cortex, cruciate (but see *K. microphylla*).

Those outer medullary cells which develop long processes are known as stellate, arachnoid or ganglionic cells; their function may be to strengthen the blade. They are well-developed in *Kallymenia, Cryptonemia* and *Halymenia*, less so in *Grateloupia*, and are diagnostic of these genera, q.v., in the British Isles. Since they lie in the plane of the blade, they are difficult to observe in section and are best seen in a squash preparation after treatment with dilute acid. Some of the characters previously suggested for distinguishing species and even genera have been found to be invalid. For example, the texture and thickness of a blade vary considerably according to its age, and the degree of protrusion of the cystocarp depends on the thickness and compactness of the blade. Codomier (1969, 1971, 1974a) found that in the Mediterranean *K. reniformis* never becomes as dark brown and cartilaginous as *K. microphylla* and he also described differences in the stellate cells. This character needs to be studied further in the British Isles and is indicated by * in the key. The genus *Meredithia* was separated by J. Agardh (1892) from *Kallymenia* by having larger cystocarps and a thicker, more fleshy, less gelatinous blade. Since the development of the cystocarps is closely similar in both genera, *Meredithia* is not accepted here and the single British species is replaced in *Kallymenia* as *K. reniformis* (but see Guiry & Maggs, 1982).

Among the unidentified collections of sublittoral foliose plants with stellate cells in the medulla (see note under *Halymenia*) are some resembling Mediterranean species of *Kallymenia* (Codomier, 1971) such as *K. requienii* (J. Ag.) J. Ag., *K. patens* Feldm. and *K. feldmannii* Codomier.

During the early years of preparation of this work some specimens of *Schizymenia dubyi* (Chauv. ex Duby) J. Ag. in which gland cells could not be seen (see 1(1) p. 177) were identified as *K. reniformis*, e.g. from Wexford (Norton, 1970).

KEY TO SPECIES

Blade not more than 50 mm broad, rigidly undulate and contorted, cannot be
pushed flat when fresh, becoming palmately lobed; cystocarps in small groups
(<8) near blade apices, up to 2 mm diameter; tetrasporangia unknown in
blade in British Isles; (*stellate cells in centre of blade pale yellow, arms 100 μm
long) .... *K. microphylla*
Blade up to 250 mm broad, softer, only margins undulate, easily pushed flat when
fresh, sometimes split, sometimes with circular marginal proliferations; cystocarps in large numbers in patches covering ¼–¾ of blade, up to 1 mm diameter;
tetrasporangia common in blades; (* stellate cells in centre of blade deep yellow,
arms to 300 μm long) .... *K. reniformis*

Kallymenia microphylla J. Agardh (1851), p. 288.

Lectotype: LD (Herb. Alg. Agardh. 24284). Devon (Torquay).

Meredithia microphylla (J. Agardh) J. Agardh (1892), p. 74.

Mature fronds of gametangial phase with a branched stipe up to 50 mm long and 1·5 mm in
diameter showing scars of old fronds, and giving rise to a series of more or less alternately
arranged blades; blades strongly concave and auricled at the base, the older ones enveloping
the younger, up to 20 mm long and 35(50) mm broad, young blades thin (50 μm), rose-red,
older blades thicker (to 400 μm) and purplish red when fresh, drying darker or black,
becoming palmately lobed and then rigidly undulate and variously contorted.

Structure multiaxial, medulla with filaments 5–10 μm in diameter and well-developed
stellate cells, cortex of several layers of cells, outermost 4–9 μm in surface view.

Spermatangia unknown; cystocarps very large, up to 2 mm, in small groups of up to 8 near
apex of blade, gonimoblast producing groups of carposporangia 15×11 μm elevating the
thickened cortex on one or both sides of the blade, spores released by gelatinisation of an
area of wall; tetrasporangia unknown in erect fronds in the British Isles, reputedly cruciate.

Thallus of tetrasporangial phase filamentous and closely similar to the *Hymenoclonium*

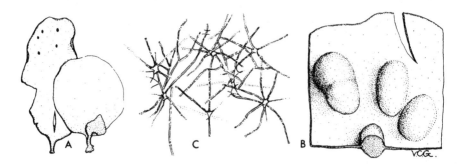

Fig. 13   *Kallymenia microphylla*
A. Habit of Type (undated) × 1; B. Part of blade with cystocarps (Aug.) × 8; C. Squash
preparation of stellate medullary filaments (July) × 80.
(T.S. as in Fig. 14)

phase known for several other genera of the Cryptonemiales, see 1(1), p. 156; tetrasporangia 18–33×10–13 μm, spores irregularly zonately arranged.

Epilithic, under overhangs and on vertical rock faces and steep slopes; sublittoral to 14 m.

Recorded from Isle of Wight, Dorset, Devon, Cornwall, Pembroke, Anglesey; in Ireland from Wexford, Clare, Galway and Donegal; probably widely distributed in suitable habitats in the sublittoral; the report for St Kilda has not been confirmed.

British Isles to Morocco; Mediterranean; Canary Isles.

Perennial, stipe increasing in length each year, plants bearing blades of different ages, the oldest as remnants. Codomier (1969) gives details of the development of plants from year to year in the Mediterranean but similar observations have not been made in the British Isles. Spermatangia unknown; cystocarps recorded for November and March–May, tetrasporangia not recorded in the field for the British Isles.

According to Codomier (1969), the young blades are rose-red and rather soft in spring, becoming much thicker, brown and cartilaginous later.

This species is said by divers to be very distinct from *K. reniformis* under water and it usually grows in more shaded situations. Since no anatomical differences have yet been found to separate the two species, herbarium specimens without cystocarps are difficult to identify with certainty and so it has not been possible to evaluate most previous reports for the British Isles. Codomier (1969, 1971) gave detailed descriptions of Mediterranean plants which seem to agree with British Isles material in most respects except that the cortical cells of the younger blades are larger (up to 15 μm) in surface view. He also found (1973) that the carpospores develop differently from those of other species of *Kallymenia*, producing a filamentous phase which he considered resembled illustrations of *Rhodochorton hauckii* (Schiffn.) Hamel, also described for the Mediterranean. His illustration shows a plant similar to the *Hymenoclonium* phases of other algae such as *Acrosymphyton, Bonnemaisonia, Pikea* and *Schimmelmannia*, (see 1(1), p. 156). Guiry & Maggs (1982) have recently found that cultured carpospores of Irish material germinated to form a similar *Hymenoclonium* phase which produced irregularly zonate tetrasporangia 18–33 × 10–13 μm.

Older plants are often covered with epiphytic coralline algae, bryozoa and sponges.

Kallymenia reniformis (Turner) J. Agardh (1842), p. 99.

Lectotype: original illustration (Turner, 1809, pl. 113, fig. b; see Norris, 1957) in the absence of specimens. Devon (Budleigh Salterton).

Fucus reniformis Turner (1809), p. 110.
Kallymenia larterae Holmes (1907), p. 85, pl. 484B (as *Callymenia Larteriae*).

Erect fronds arising from a small attachment disc, with a simple or branched terete stipe 3–15 mm in length and 1–2 mm in diameter which expands abruptly into a thin but strong blade; blade translucent, mucilaginous only when young, often broader than long, up to 120 mm long and 180 mm broad, up to 375 μm thick, rose to dark purplish or brownish red, base cuneate to reniform, margins often undulate and in older fronds developing proliferations resembling the primary frond, also regenerating similarly from old, worn blades which thus appear much divided.

Structure multiaxial; medulla loose, consisting of branched elongated filaments interspersed with rhizoids and large rounded cells 20–40 μm in diameter, some of which develop long processes lying mainly in the plane of the blade (stellate cells), cortex with short radial rows of smaller cells about 6–9 μm in diameter in surface view.

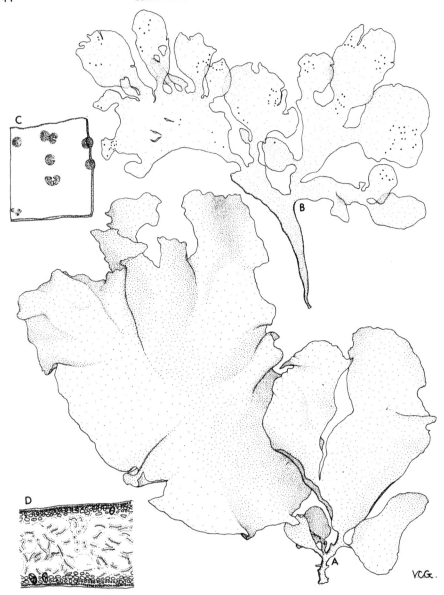

Fig. 14   *Kallymenia reniformis*
A. Habit (Sep.) × 1; B. Habit of old regenerating plant with cystocarps (Sep.) × 1; C.
Part of blade with cystocarps (Sep.) × 8; D. T.S. blade with tetrasporangia (Sep.) × 80.
(Stellate medullary filaments as in Fig. 13)

Spermatangia unknown; gonimoblasts with a compound lobed fusion cell, cystocarps to 1(2) mm in diameter, with an elevated and folded cortex, without a pore, occurring in patches of varying size in the upper two-thirds of the blade, conspicuous to the naked eye, most cells forming carposporangia 15–18 μm in diameter with a few enveloping filaments; tetrasporangia scattered, developing from an inner cortical cell, 30–33×15–21 μm, with cruciately arranged spores.

Epilithic and on *Laminaria* stipes, extending from deep pools in the upper sublittoral to at least 27 m.

Eastwards to the Isle of Wight and northwards to Shetland; reports for Norfolk and Northumberland not confirmed; in Ireland eastwards to Waterford, northwards to Galway; Antrim.

British Isles to Morocco; Mediterranean (Algeria); Canary Isles. Records from USA (Massachusetts), S. Africa and Pacific not verified.

Perennial but with new blades developing annually; cystocarps occurring from July–November; these occasionally persist over the winter and become very large (2 mm) (Irvine *et al.*, 1975); tetrasporangia recorded from May–November.

Variation in primary blade appearance may be due to ecological factors such as degree of exposure to wave action; plants with thin, undulate blades have been called var. *undulata* J. Ag. whilst those with proliferations have been called var. *ferrarii* J. Ag. and *Fucus reniformis* var. *tenuior* Turn. Old plants with small new blades arising from the margins of a worn and lacerated primary blade have been distinguished as *K. larterae* Holmes.

The 'pore' in the cystocarp illustrated by Fritsch (1945) and Kylin (1956) is probably only an invagination of the cortex. Hommersand & Ott (1970) described an ostiolate pericarp but their figure shows a rather indefinite pore probably representing gelatinization of the cystocarp wall before spore release.

CHOREOCOLACACEAE Sturch

CHOREOCOLACACEAE Sturch (1926), p. 602 [as Choreocolaceae].

Thalli with prostrate endophytic (?parasitic) filaments ramifying within the host and comparatively small, cushion-like or lobed erect fronds externally in which reproductive structures are borne; multiaxial, medulla pseudoparenchymatous, cortex compact; supporting cell functioning as or cutting off auxiliary cell (*Choreocolax, Harveyella*), or auxiliary cell branches remote (*Holmsella*), gonimoblast developing inwards or outwards, fusion cell present or absent, carposporangia terminal or in rows, cystocarps immersed or somewhat protruding; tetrasporangia scattered in cortex, cruciate.

Three genera with British Isles representatives, all reputedly parasitic (see 1(1) p. 49; Evans *et al.*, 1978, 1981), have been assigned to this family: *Choreocolax, Harveyella* and *Holmsella*. There are differences in the details of carposporophyte development (Sturch, 1924) in these genera although in other respects there is a close resemblance between them. Kylin (1956) felt that the differences might warrant the exclusion of *Holmsella* from the Choreocolacaceae; the formation of the auxiliary cell after fertilization suggests an affinity with the Ceramiales; see Levring (1935) and Norris (1957).

CHOREOCOLAX Reinsch

CHOREOCOLAX Reinsch (1875), p. 61.

Type species: *C. polysiphoniae* Reinsch (1875), p. 61.

Thallus parasitic on *Polysiphonia* in British Isles; consisting of filaments penetrating into host tissue and comparatively small cushions externally; structure multiaxial, medulla pseudoparenchymatous, surrounded by a cortical layer of smaller cells.

Gametangial plants dioecious; spermatangia covering whole male cushion; carpogonial branches 4-celled, supporting cell functioning as, or cutting off, auxiliary cell after fertilization, gonimoblast developing in medulla, with an irregular fusion cell and closely packed filaments with terminal carposporangia, enveloping filaments and pore present; tetrasporangia embedded in cortex, scattered, cruciate.

One species in the British Isles:

Choreocolax polysiphoniae Reinsch (1875), p. 61.

Lectotype: HBG ? (not seen). Atlantic coast of N. America.

Thallus parasitic on *Polysiphonia lanosa* (L.) Tandy, cushion-like, lobed, up to 1 mm in extent with an extensive filamentous basal portion ramifying between the cells of the host and forming numerous pit connections with them; colour pale brownish or colourless owing to the reduction of chloroplasts.

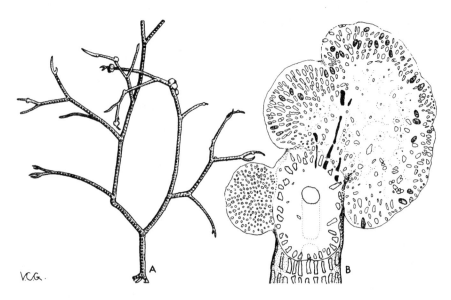

VCG.

Fig. 15   *Choreocolax polysiphoniae*
A. Several plants on *Polysiphonia lanosa* (June) × 6; B. V.S. tetrasporangial plant on host (June) × 80.

Structure multiaxial; medullary cells about 25 μm in diameter, cortical cells 6–10 μm in diameter, sheath surrounding the whole remaining thin.

Gametangial plants dioecious; spermatangia developing in clusters from the superficial cortical layer, the cells of which function as mother cells, spermatia 3×5 μm; cystocarps embedded in the medulla, 5–8 per cushion, each with a pore and enveloping filaments, carposporangia borne in pairs towards the centre of the cystocarp, 30–40×10–18 μm; tetrasporangia scattered in the cortex, 30–80×15–30 μm, spores cruciately arranged.

Reported by Sturch (1926) to be frequent wherever the host occurs. Martin (pers. comm.) found that *C. polysiphoniae* apparently thrives in areas with much sediment and suggested that this abraded the surface of the host; large populations of *P. lanosa* without the parasite occurred in areas on the west of Anglesey where the water is relatively clear.

Recorded for scattered localities throughout the British Isles.

Faroes; North Russia (Barents Sea) to N. W. Spain; Canada (Newfoundland) to USA (Connecticut); USA (Alaska) to Mexico (Baha California).

Cushions have been recorded throughout the year, though they are scarce in November–December and the species apparently overwinters by means of filaments embedded in the host.

Tetrasporangia occur throughout the year and gametangia in spring and summer. Sturch (1926) collected plants every fortnight and found that all mature plants were fertile, with tetrasporangial, male and female in equal proportions. Kugrens & West (1975) reported that in California the cushions were absent from November to January and very few collections were fertile.

Polysiphonia plants infected by *C. polysiphoniae* are less vigorous (Richards, 1891) but Kugrens & West found no bacteria or viruses in either entity to account for this condition. Host disruption and translocation of substances from host to algal parasite are discussed by Evans *et al.* (1978).

Apart from a possible specimen on *Polysiphonia elongata* (Huds.) Spreng. collected by Holmes from Studland in April 1889 (BM, Goff, pers. comm.), the only recorded host for *C. polysiphoniae* in the Atlantic is *P. lanosa* (L.) Tandy. Zinova (1970) has reported it on *P. urceolata* (Lightf. ex Dillw.) Grev. in North Russia. In the Pacific the species is reported growing on a range of hosts including species of *Polysiphonia, Pterosiphonia* and *Pterochondria*.

Excluded species

Choreocolax cystoclonii Kylin (1907) p. 127.
Choreocolax tumidus Reinsch (1875) p. 65.

The original material of these entities has been shown to consist of bacterial galls on *Cystoclonium purpureum* (Huds.) Batt. and a species of *Ceramium*, respectively. Subsequent reports were probably based on similar material.

HARVEYELLA Schmitz & Reinke in Reinke

HARVEYELLA Schmitz & Reinke in Reinke (1889), p. 28.

Type species: *H. mirabilis* (Reinsch) Schmitz & Reinke in Reinke (1889), p. 28.

Thallus reputedly parasitic on *Rhodomela* and *Odonthalia*, with a lower portion penetrating into host tissue and small protruding cushions externally; structure multiaxial, medulla pseudoparenchymatous, surrounded by a cortical layer of smaller cells.

Gametangial plants dioecious; spermatangia covering whole male cushion; carpogonial branches 4-celled, supporting cell functioning as auxiliary cell, gonimoblast with a large fusion cell, developing inwards before producing filaments with terminal sporangia outwards, enveloping filaments and pores present; tetrasporangia embedded in cortex, scattered, cruciate.

It was at first thought that the auxiliary cell was cut off from the supporting cell, probably after fertilization, suggesting an affinity with the Ceramiales. Later workers have shown that the supporting cell functions as the auxiliary cell. Goff & Cole (1975) found that it was not possible to determine the phyletic origin of the reproductive filaments, so that the assignment of *Harveyella* to an order is at present subjective.

One species in the British Isles:

Harveyella mirabilis (Reinsch) Schmitz & Reinke in Reinke (1889), p. 28.

Lectotype: HBG? (not seen). Isotypes: Rabenhorst's Algen Europa's no. 1878. Sweden (Bohusland).

Choreocolax mirabilis Reinsch (1875), p. 63.
Choreocolax odonthaliae Levring (1935), p. 55.

Thallus reputedly parasitic on *Rhodomela confervoides* (Huds.) Silva, cushionlike, up to 2 mm in extent with a filamentous basal portion ramifying between the cells of the host and forming pit connections with them, colour pale brown or whitish owing to the reduction or absence of chloroplasts. Structure multiaxial; medullary cells in vertical section irregular, $28–40 \times 9–11$ μm, cortical cells $22–33 \times 13–20$ μm, sheath surrounding the whole remaining thin.

Gametangial plants dioecious; spermatangia developing in clusters from the outer cortex the cells of which function as mother cells, spermatangia in rows cut off by alternating oblique walls, spermatia *c.* 4 μm; gonimoblast filaments developing in groups between elongated vegetative cells, producing terminal carposporangia outwards, a single cystocarp occupying most of a cushion, carpospores $17–30 \times 10–15$ μm, released through small pores; tetrasporangia scattered in the cortex, $25–45 \times 15–20$ μm, tetraspores cruciately arranged.

Goff & Cole (1976a) found that, in the eastern Pacific, the species most commonly occurs in wave-sheltered areas, and usually infects older parts of the host. Plants in the upper littoral, where the water temperature reaches 18–25°C, produced very few spores. Known to a depth of 8 m in the British Isles, but reported by Rosenvinge (1931) from 26 m in Denmark.

Recorded for localities scattered throughout the British Isles.

Iceland; Faroes; North Russia (Barents Sea) to France; Baltic; E. Greenland; E. Canada (Labrador) to USA (Rhode Is.); North Pacific.

Spermatia recorded for September to February; cystocarps in January–February and August; tetrasporangia in January, July and September. Sturch (1924) observed that in Plymouth spores were liberated between January and March. These produced filaments within the host, the cortex of the host increased and swellings became visible externally in October, after which the cushion grew rapidly, bursting the host tissues, and produced

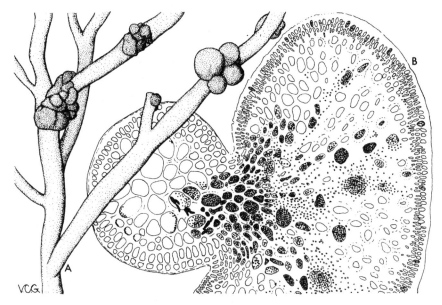

Fig. 16   *Harveyella mirabilis*
A. Several plants on *Rhodomela confervoides* (Jan.) × 8; B. V.S. tetrasporangial plant
showing filaments of parasite within host (June) × 80.

gametangia in late October and November. Goff & Cole (1973, 1976a) found that, in the
eastern Pacific, the species had a *Polysiphonia*-type life history, passing through a complete
cycle annually. They attempted to correlate reproductive periodicity with environmental
factors, particularly water temperature. They suggested that gametangia are produced in
summer only in areas where the maximum temperature is below about 12°C; elsewhere (e.g.
Plymouth) they are produced in winter. Some of the records from the British Isles appear to
be anomalous, however, and more data are required, especially as Sturch (1924) found
differences in behaviour between littoral and sublittoral plants.
    Goff & Cole (1976) found that successful germination appeared to depend on the entry of a
spore into a lesion of a host plant such as that caused by a grazing animal. Goff (1976) found
that penetrating filaments of *Harveyella* damaged the host tissue, causing chloroplast
distortion, death of medullary cells and hollow branches. The secondary pit-connections
between parasite and host have been studied by Peyrière (1981).
    It was thought for a long time that, in the North Atlantic, the species occurred only on
Rhodomela spp. Edelstein & McLachlan (1977) have shown, however, that *Choreocolax
odonthaliae* Levring is synonymous with *H. mirabilis* and so the species is now known to
occur on both *Rhodomela* and *Odonthalia*, as it does in the North Pacific. It has not so far
been recorded on *Odonthalia* in the British Isles.

HOLMSELLA Sturch

HOLMSELLA Sturch (1926), p. 603.

Type species: *H. pachyderma* (Reinsch) Sturch (1926), p. 604.

Thallus parasitic on *Gracilaria verrucosa* (Huds.) Papenf.; with a lower portion penetrating into host tissue and comparatively small cushions externally; structure multiaxial with a pseudoparenchymatous medulla and a cortical layer of smaller cells.

Gametangial plants dioecious; spermatangia covering whole male cushion; carpogonial branches usually 2-celled, borne laterally on cortical filaments, auxiliary cell branches formed after fertilization from random cortical cells, gonimoblast producing rows of carposporangia outwards; tetrasporangia lateral, scattered in cortex, cruciate.

Holmsella pachyderma (Reinsch) Sturch (1926), p. 604.

Lectotype: HBG ? (not seen). Isotypes: Hohenacker's Algae marinae no. 346. France (Arromanches, *fide* label).

Choreocolax pachydermus Reinsch (1875), p. 62.

Thallus almost spherical, up to 1 mm in diameter with a basal filamentous portion embedded in the fronds of *Gracilaria verrucosa* (Huds.) Papenf., the filaments ramifying between the cells of the host and forming pit connections with them; colour almost white owing to the reduction of chloroplasts.

Medullary cells in vertical section irregular, 16–23 μm, cortical cells smaller, isodiametric, 4–8 μm at the periphery, the whole surrounded by a translucent sheath 50–110 μm thick.

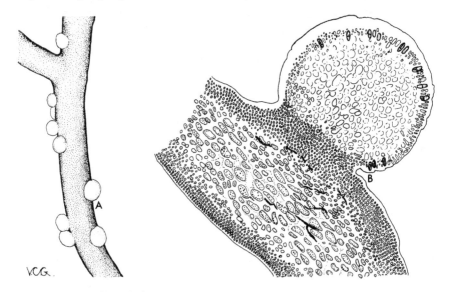

VCG.

Fig. 17   *Holmsella pachyderma*
A. Several plants on *Gracilaria verrucosa* (April) × 8; B. V.S. tetrasporangial plant showing filaments of parasite within host × 80.

Gametangial plants dioecious; spermatangia not clustered, developing evenly from outer cortical cells over large areas, spermatia cut off by transverse walls; cystocarps in the cortex, usually one per cushion, gonimoblast filaments interspersed among cortical cells some of which become radially elongated, producing carposporangia in rows outwards, carpospores 9–15×7–10 μm; tetrasporangia lateral on cortical filaments, scattered, 24–25×11–17 μm, tetraspores cruciately arranged.

Lower littoral and sublittoral to 16 m on *Gracilaria verrucosa* (Huds.) Papenf.

Southern and western shores of the British Isles, extending eastwards to Sussex, Norfolk, and northwards to the Isle of Man; Galway, Down, Wexford. Southern parts of the British Isles to N. W. Spain.

Cushions have been recorded throughout the year. Sturch (1924) reported that the parasite passes through two complete cycles annually in deep water, but only one at higher levels, between October and April, the cushions disappearing from the host plants in the littoral after spore discharge. Martin, however, (pers. comm.) has found that cushions reappear in July or earlier on Anglesey. The penetrating filaments are probably perennial and could persist in basal parts of the host buried in sand. Gametangia and tetrasporangia recorded from November–March.

Reinsch (1875) gives the type locality as Mediterranean but the specimen of Hohenacker's Algae Marinae no. 346 in BM is labelled Arromanches, Calvados; this number appears to have been a uniform gathering (Anon., 1860).

Although Sturch reported that the filaments sometimes penetrated the host cells, this was not confirmed by Evans *et al.* (1973). They did, however, demonstrate the transfer of photosynthetically fixed ^{14}C from the host to the parasite; see also Quirk & Wetherbee, 1980. The secondary pit-connections between parasite and host have been studied by Peyrière (1981).

The cushions of the parasite are easily distinguished from the cystocarps of the host (which have a prominent pore) or galls because they are smaller and pearl-like.

PEYSSONNELIACEAE

by

Linda M. Irvine

and

Christine A. Maggs*

PEYSSONNELIACEAE Denizot

PEYSSONNELIACEAE Denizot (1968), p. 86.
Squamariaceae (Zanardini) J. Agardh (1851), p. IX, nom. illeg., pro parte.
Squamarieae Zanardini (1842), p. 235, non Squammarieae Fee (1824), p. LIV.

Thallus prostrate, sometimes completely adherent to substrate, sometimes partly detached; consisting of a single basal layer of closely united branched filaments each cell of which produces a single cell above giving rise to filaments branched in a characteristic manner, some cells of basal layer also producing rhizoids below; gametangial plants mono- or dioecious, carpogonial branches amongst sterile filaments in sori, auxiliary cell branches 3–6-celled, connecting filament fusing with third cell, carposporangia in more or less branched rows; tetrasporangia amongst sterile filaments grouped in sori, cruciate. Calcification, when present, of aragonite.

Denizot (1968) included three genera (*Cruoriella, Peyssonnelia* and *Polystrata*) in his revision of this family; one of these, *Peyssonnelia,* is well-defined and represented by at least 4 species in the British Isles. *Haematocelis* has also been provisionally assigned to this family here because the vegetative thallus resembles that of some species of *Peyssonnelia.* Its taxonomic position has been discussed by Ardré (1977, 1980) and Maggs & Guiry (1982a) who found that *Haematocelis*-like crusts are implicated in the life-history of *Schizymenia* (see 1(1) p. 175) and *Sphaerococcus* (see 1(1) p. 204) respectively. DeCew & West (1981) found that, in the eastern Pacific, *Haematocelis*-like crusts are involved with species of *Farlowia* (Dumontiaceae).

HAEMATOCELIS J. Agardh

HAEMATOCELIS J. Agardh (1851), p. XII.

Type species: *H. rubens* J. Agardh (1852), p. 497.

Haematophlaea (J. Agardh) Crouan (1858), p. 73 pro parte quoad descr., non *Hildenbrandia* sect. *Haematophlaea* J. Agardh (1852), p. 495 quoad typus.

Thallus encrusting, comparatively thick, firm and cartilaginous, closely adherent to substrate, rhizoids absent; basal layer producing firmly united, curved then erect, branched filaments.

* Department of Botany University College, Galway, Republic of Ireland

Gametangial crusts unknown; tetrasporangia in immersed sori, terminal on erect filaments, zonate.

One species in the British Isles (but see Maggs & Guiry, 1982a):

Haematocelis rubens J. Agardh (1852), p. 497.

Holotype: LD (Herb. Alg. Agardh. 27625). Isotype: CO. France (Brest).

Haematophlaea crouanii (J. Agardh) Crouan (1858), p. 73 pro parte, quoad descr., non *Hildenbrandia* sect. *Haematophlaea crouanii* J. Agardh (1852), p. 495.

Thallus encrusting, up to 50 mm or more in extent and 700 μm thick and appearing three or more layered in vertical section; dark red, drying black; consisting of a basal layer of radiating prostrate filaments closely adherent to the substrate, without rhizoids, giving rise to firmly united, branched filaments at first almost prostrate, irregular, to 15 μm in diameter and frequently containing conspicuous globules 6–10 μm in diameter, becoming erect, forming a middle layer which appears stratified due to synchronous cell divisions, cells rectangular, 18–22×4–9(12) μm, becoming narrower above with shorter cells which appear rounded and 4–5 μm in surface view; thicker thalli showing growth zones.

Gametangial crusts unknown; tetrasporangia among sterile filaments in the uppermost layer, 48–80×12–22 μm, spores zonately arranged; borne on filaments often noticeably thicker than vegetative filaments (to 12 μm); sori becoming rather mucilaginous.

Epilithic and epiphytic on *Laminaria* stipes and holdfasts; upper sublittoral in conditions of moderate exposure to wave action.

Recorded for Cornwall, Devon, Dorset, Northumberland and Wexford, but probably occurring throughout the British Isles. British Isles southwards to Morocco; Sicily; Tristan da Cunha; California.

Little is known about the periodicity of growth but plants are probably perennial since the thicker thalli show marked horizontal growth zones; gametangia unknown, tetrasporangia recorded for December, January and August, sometimes becoming embedded with the spores remaining unshed.

Ardré (1977) and Sciuto *et al*. (1979) have found evidence from field studies and chemistry, respectively, which appears to implicate this species in the life history of *Schizymenia dubyi* (Chauv. ex Duby) J. Ag. (see 1(1), p. 176). Ardré discovered that the basal crust of *S. dubyi* is similar to *H. rubens* in organization, structure and cell dimensions, and in two specimens she found tetrasporangia continuous or subjacent to the point of emergence of the erect frond from the crust. More recently, cultured tetraspores from a *Haematocelis*-like crust have produced erect fronds of *Sphaerococcus coronopifolius* Stackh. see 1(1), p. 204 (Maggs & Guiry, 1982a). The crust has smaller tetrasporangia ((29)35–42(52)×10–12 μm) than *H. rubens* and the surface shows white spots which appear to be intercellular oil globules; it seems to be identical with *Haematocelis fissurata* Crouan frat. (= *Ethelia fissurata* (Crouan frat.) Denizot), and has been found in the sublittoral (3–24 m) in 3 localities in Ireland and 2 in the Hebrides, Scotland. Earlier reports of this species for the British Isles were based on specimens of the crust of *Dumontia contorta* (S. G. Gmel.) Rupr., q.v. (Brebner, 1895). See also DeCew & West (1981).

PEYSSONNELIA Decaisne

PEYSSONNELIA Decaisne (1841), p. 168.

Type species: *P. squamaria* (S. G. Gmelin) Decaisne (1841), p. 168.

Squamaria Zanardini (1841), p. 133, nom. illeg., non *Squamaria* Ludwig (1757), p. 120.
Nardoa Zanardini (1844), p. 37.
Gymnosorus Trevisan (1848), p. 108.
Lithymenia Zanardini (1863), p. 127.
?*Haematostagon* Strömfelt (1886), p. 173.

Thallus prostrate and encrusting, not usually mucilaginous, thin to comparatively thick, closely attached to the substrate by rhizoids which arise from the whole undersurface but degree of adherence variable, margins sometimes curving or rolling upwards, especially when dry, underside with a layer of extracellular calcification (aragonite); consisting of a single basal layer of branched filaments which are either little branched and more or less parallel or much branched to form a polyflabellate layer; each basal cell cutting off a single cell from which closely united branched ascending filaments arise.

Gametangial plants mono- or dioecious; gametangia in superficial or immersed sori, spermatangia produced in dense clusters within a wall, with or without accompanying sterile filaments; carpogonial branches 3–6-celled, auxiliary cell branches remote, 3–6-celled, gonimoblasts developing outwards, consisting of comparatively few cells, almost all becoming carposporangia more or less arranged in short rows; tetrasporangial sori superficial or immersed, tetrasporangia terminal or apparently lateral, cruciate, accompanied by sterile filaments.

Elsewhere the genus exhibits a wider range of form (Maggs & Irvine, 1983, in press): some species have much branched or lobed fronds which adhere at only one point and loose-lying populations have been recorded; also the thallus itself is sometimes calcified or contains cystoliths.

Plants of the genus *Peyssonnelia* are difficult to collect and study not only because they are usually found encrusting rock or prostrate coralline algae but also because they have a more complex anatomy than most other encrusting genera and are partially calcified. Some characters previously used to distinguish species are invalid (e.g. Newton, 1931), particularly those relating to the length and diameter of the cells in the ascending filaments. The strict branching pattern of these filaments is such that the filament diameter is halved above a branch (see Newton, 1931 Fig. 183); the cell length is variable and probably related to the rate of growth. The anatomical features were not appreciated until comparatively recently (see Denizot, 1968) and so many previous reports for the British Isles are erroneous. Although considerable effort has been made to re-evaluate them, the specimens on which they were based are often either lost or fragmentary and the situation is still unsatisfactory. Furthermore, plants of different species often occur together on the same substrate and may even become confluent; confusion has been further increased when gatherings consisting of mixtures have subsequently been divided and dispersed. Species of this genus are notoriously variable and specialists advise against attempting to identify single sterile specimens (see Denizot, 1968; Boudouresque & Denizot, 1975).

The present treatment is based mainly on recent studies of newly collected material together with an examination of selected type and other critical specimens. When studying the vegetative structure it is essential to use only carefully orientated sections, usually radial vertical ones (RVS, see Fig. 18). Random sections are useless and misleading.

The degree of adherence provides a useful character but can be meaningfully assessed only

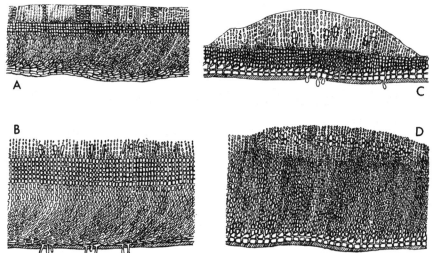

Fig. 18   *Haematocelis* sp. and *Peyssonnelia* spp.
A. R.V.S. *H. rubens* tetrasporangial crust (Jan.) × 95; B. R.V.S. *P. atropurpurea* crust
with immersed tetrasporangial sorus (Jan.) × 95; C. R.V.S. *P. dubyi* crust with
superficial tetrasporangial sorus (Mar.) × 95; D. R. V. S. *P. immersa* crust with immersed
sorus showing superposed carposporangia resembling tetrasporangia (Sep.) × 95
(rhizoids not shown). Sori of *P. harveyana* are similar to those of *P. dubyi*. Hatching
indicates hypobasal calcification.

in relatively recently collected plants. Most herbarium specimens of any species of *Peyssonnelia* gradually become detached from the substrate after a few months. In an investigation of
Mediterranean species, Boudouresque & Denizot used some new taxonomic criteria such as
the growth and slope of the ascending filaments and form and segmentation of the rhizoids;
Marcot, Boudouresque & Verlaque (1977) subsequently used several characters associated
with the tetrasporangial sori. Maggs & Irvine (1983) have drawn attention to the difference
between the superficial and immersed types of sorus development. In the former, the sorus is
very mucilaginous and consists of modified, more or less colourless filaments on or between
which the reproductive bodies are borne. In the latter, the filaments are pigmented and little
modified from the vegetative state, especially the upper cells which form a roof to the
developing sorus; these sori become mucilaginous and raised only at maturity. Other papers
containing information relevant to species occurring in the British Isles include Boudouresque & Ardré (1971), Bressan (1972), Belsher & Marcot (1975), Marcot & Boudouresque
(1976), and Verlaque (1978).

Four species are recognised here; all are quite distinct, at least when fertile. Three belong
to the subgroup with a '*P. rubra*'-type anatomy (Denizot 1968) which contrasts markedly
with that of the remaining species, *P. atropurpurea*.

Three species previously recorded for the British Isles are not included: *P. rosenvingii*
Schmitz in Rosenvinge (1893, p. 782), *P. rubra* Greville (1826, p. 340) and *P. rupestris*
Crouan frat., (1867, p. 148). *P. rosenvingii* was originally described from W. Greenland and
has been widely recorded for Arctic Europe, the British Isles and from Arctic Canada to

Brazil. Denizot (1968) considered *P. rosenvingii* to be doubtfully distinct from *P. harveyana* and the specimens supporting British records bear a close resemblance to this species. It is not regarded as a synonym here, pending further investigation of material from elsewhere.

P. rubra is a distinctive Mediterranean species in which certain enlarged basal cells contain special calcified bodies called cystoliths. Specimens from the British Isles identified as *P. rubra* do not possess cystoliths and belong to either *P. harveyana* or *P. dubyi*. Bressan (1972) reported *P. rubra* from France (Baie de Morlaix).

P. rupestris was shown by Denizot (1968), to be a synonym of *Rhodophysema elegans* (Crouan frat. ex J. Ag.) Dixon, q.v.

KEY TO SPECIES

1   In RVS, ascending filaments at an angle of less than 30° to basal layer, arising from a cell cut off from anterior end of basal filament cell; upper surface usually showing concentric zones .... *P. atropurpurea*
    In RVS, ascending filaments at an angle of more than 60° to basal layer, arising from a cell cut off from whole upper surface of basal filament cell; upper surface not showing concentric zones 2
2   Basal layer polyflabellate, showing a network of branching and anastomosing filaments when viewed from below; upper surface distinctly wrinkled when dry, especially near margins .... *P. dubyi*
    Basal layer not polyflabellate, showing radial or curved rows of parallel basal filaments when viewed from below, upper surface smooth, at least near margins 3
3   Upper surface distinctly radially striate; margins free, leaving a gap when growing over rough substrates .... *P. harveyana*
    Upper surface with sinuous striations; margins closely adpressed, even to rough substrates .... *P. immersa*

Peyssonnelia atropurpurea Crouan frat. (1867), p. 148.

Lectotype: CO. France (Brest). Isotypes: Algues Marines du Finistère no. 237 (see Denizot, 1968 fig. 93).

Peyssonnelia atropurpurea Crouan frat. in Le Jolis (1863), p. 129, nom. nud.
Haematocelis schousboei J. Agardh (1851), p. XII, (1852) p. 498, pro parte, excl. typ.

Crusts adherent in the centre, loose at the margins which roll upwards on drying, entire or lobed, especially on uneven substrates, to 100 mm or more in extent and to about 500 μm thick, carmine to dark purplish red, glossy when dry and usually showing concentric rings; rhizoids unicellular, arising from anterior end of basal cells, 55–110 μm long, usually numerous, 12–15 μm in diameter; basal filaments in radial rows, appearing parallel, each cell elongated and partially curved upwards, cutting off a single cell from anterior part of upper surface; in radial vertical section ascending filaments inclined towards margin at a very narrow angle (less than 30°) but becoming erect above in older thalli, and about 8–10 μm in diameter, cells (1)2–3(5) times as long, growth zones commonly seen, calcification hypo-basal, up to at least 110 μm thick.

Gametangial plants reputedly dioecious; spermatangial sori immersed, pale, spermatangia terminal; carpogonial sori immersed, carpogonial and auxiliary cell branches borne laterally

8–13 cells below apex of little modified, pigmented filaments which are sometimes pseudo-dichotomous, most cells elongated, the uppermost broad and fused to form a roof, mature carposporophytic sori 130–210 μm deep, little raised even when mature, gonimoblast arising from connecting filament, composed of 4–8 cells most of which become carposporangia 10–30 μm in diameter; tetrasporangial sori to 100 μm deep, not mucilaginous, extensive, immersed, little raised even when mature, tetrasporangia terminal with a conspicuous stalk cell among little-branched clubshaped pigmented filaments 50–90 μm long composed of 4–6 cells when mature, tetrasporangia 40–80×20–27 μm with cruciately arranged spores; repeated generation of tetrasporangia not known.

Epilithic and epiphytic, mainly on crustose corallines, sometimes overgrowing other algae, sand and debris, in shady places; upper sublittoral to at least 13 m.

Eastwards to Dorset, northwards to Lundy; Isle of Man; Wexford, Cork, Galway; Channel Isles.

British Isles to Morocco (Tangier).

Perennial, spermatangial plants not recorded in British Isles; cystocarps recorded for January; most of them were empty, indicating that carposporophyte development occurred earlier; tetrasporangia recorded from November–April. The spermatangia described here are from plants collected in France (Roscoff) in October; Belsher & Marcot (1975) gave a detailed account of this population and recorded gametangia and tetrasporangia in November. In Portugal Ardré (1970) found cystocarps from February–October and tetrasporangia in June.

Anomalies in structure are frequently seen, especially in plants growing over irregular substrates, and the fronds are then more subdivided.

This species can be confused with *Haematocelis rubens* J. Ag. from which it differs by having cruciate tetrasporangia and a calcified undersurface pierced by rhizoids.

Care is needed in the examination of thick specimens because the very oblique primary growth of the ascending filaments can be obscured by a considerable depth of erect secondary growth.

Peyssonnelia dubyi Crouan frat. (1844), p. 368.

Holotype: CO. Isotypes: PC, AHFH (54107, slides 1573–4) (see Dawson 1952). France (Banc de St. Marc, Brest).

Cruoriella dubyi (Crouan frat.) Schmitz (1889), p. 454 (reprint p. 20).

Crusts closely adherent, lobed, to 50 mm or more in extent and to about 400 μm thick, dark red, sometimes purplish or brownish, wrinkled when dry and usually glossy, older crusts sometimes scaly; rhizoids short, to 40 μm long, usually arising from anterior end of basal cell, thin-walled, 8–10 μm in diameter; basal filaments not in radial rows, but producing a polyflabellate layer, each cell cutting off from whole upper surface a single cell which is boot-shaped in radial vertical section and gives rise to compact branched filaments inclined towards margin at an angle of more than 60° to the basal layer becoming erect above and about 10 μm in diameter, cells rather shorter to rather longer than broad, occasionally twice as long; calcification hypobasal, to 80 μm thick. Secondary growth occurring by overgrowth, the thallus appearing horizontally split, with rhizoids and calcification developed secondarily from the base of the upper layer.

Gametangial plants usually monoecious; gametangia in superficial mucilaginous sori, spermatangia colourless, borne in terminal groups; carpogonia in same or separate sori to

Fig. 19   *Peyssonnelia* spp. Stylized drawings of dried specimens.
A. Habit of *P. atropurpurea* plant growing over filamentous débris, showing strongly
upturned margin and concentric zones (Jan.) × 4; B. Habit of *P. dubyi* plant on
encrusting coralline algae, with many superficial secondary outgrowths (May) × 4; C.
Habit of *P. harveyana* plant growing on shell and worm tube, showing radial striations
(June) × 4; D. Habit of *P. immersa* plant on shell, showing very close attachment to
substrate (Jan.) × 4.

150 μm deep and 1 mm in diameter, carpogonial and auxiliary cell branches 4–5 celled, borne laterally on basal or second cell of an 8-celled filament, gonimoblast composed of 8–18 cells all becoming carposporangia in rows of 2–5, 30–40 μm, including thick wall; tetrasporangial sori superficial, mucilaginous, 1–2 mm in diameter and to 100 μm deep, scattered over thallus surface; tetrasporangia terminal or lateral, 44–80 × 17–39 μm, including thick wall, with cruciately arranged spores among unbranched, colourless filaments whose cells elongate as the sporangia enlarge, repeated generation of tetrasporangia not known.

Epilithic, epiphytic on encrusting corallines, *Laminaria* holdfasts and occasionally *Graciaria verrucosa* etc, and on shells; in shallow pools in lower littoral, on pebbles in sheltered areas, sublittoral to 16 m tolerating a wide range of exposure to currents and wave action.

Generally distributed throughout the British Isles.

Norway (Trondelag) to Portugal; Mediterranean; Cape Verde Isles; Western Baltic.

Perennial; spermatangia recorded for January, February, July, August and November, cystocarps and tetrasporangia throughout the year with a peak in August–September.

There appears to be little form variation and the species is comparatively easy to recognize, even when growing over an irregular substrate.

This species has sometimes been mistakenly considered the type of the genus *Cruoriella*. One of the striking features of *P. dubyi* is the polyflabellate nature of its basal layer and this came to be thought of as diagnostic of the genus *Cruoriella*. Denizot (1968) re-established *C. armorica* Crouan frat. as the type of *Cruoriella*, however, and pointed out that it differs from *P. dubyi* because the thallus is loose and mucilaginous throughout, using this feature to redefine the genus. This feature is not shared by *P. dubyi*, although it closely resembles *C. armorica* in other respects. *C. armorica* was originally described from northern France and may also occur in the British Isles. (Found by Maggs, Freamhainn & Guiry, 1983).

According to Rosenvinge (1917), secondary growth can lead to the superposition of fronds either by overlapping or by horizontal splitting, together with regeneration of the basal layer, rhizoids and calcification; see also Printz (1926) who thought it was a direct consequence of regeneration after reproduction. This phenomenon is frequently seen and requires further study.

From Rosenvinge's original (1917) description, *P. codana* (Rosenv.) Denizot is very similar to *P. dubyi* from which it appears to differ only by its brighter colour and smaller carpospores. *P. codana* has been recorded for the Mediterranean by Verlaque (1978) but his material differs from the original by having prominent main axial filaments in the polyflabellate basal layer. Marcot, Boudouresque & Verlaque (1977) report that the tetrasporangia are inserted laterally rather than terminally as in *P. dubyi*, but in the British Isles *P. dubyi* can have laterally or terminally borne tetrasporangia in the same sorus.

Peyssonnelia harveyana J. Agardh (1852), p. 501.

Holotype: LD (Herb. Alg. Agardh. 27642, see Marcot & Boudouresque, 1976).

Crusts very adherent except at margins which roll up on drying, entire or lobed especially on irregular substrates, surface often verrucose, to 50 mm or more in extent and up to 300(425) μm thick, dark to bright red, drying brown or purplish, matt or glossy, usually conspicuously radially striate; rhizoids to 100 μm long, thin-walled, 8–18 μm in diameter and usually arising from central depression in base of basal cell; basal filaments in parallel radial (or occasionally rather sinuous) rows, more irregular in older parts and on uneven substrates, each cell cutting off from whole upper surface a single cell which is boot-shaped in radial vertical section and gives rise to compact branched filaments inclined towards margin at an angle of more than 60° to the basal layer, becoming erect and about 10 μm in diameter above, cells of filaments usually 1–2 times longer than broad below, shorter and more irregular in length above in thicker thalli, which often show growth zones; calcification hypobasal, up to 150 μm thick. Gametangial plants monoecious, gametangia in superficial mucilaginous sori which are round or elongated, often confluent, up to 1·5 mm and 180 μm thick; spermatangia in terminal groups; carpogonia in same or separate sori, sterile filaments colourless, simple, with 5–9 cells and 60–127 μm long, gonimoblasts composed of 3 or more rows of 2–3 superposed carposporangia, 25–60 μm; tetrasporangial sori forming confluent raised spots, tetrasporangia with an inconspicuous stalk cell, terminal, 100–130×30–48 μm with cruciately arranged spores among simple or pseudodichotomous, colourless sterile filaments composed of 4–6 cells, 100–150 μm long, repeated generation of tetrasporangia not known.

Epilithic, epiphytic on coralline algae and *Laminaria* holdfasts and on shells, upper sublittoral to 12 m: in habitats with at least some current flow; tolerant of wave action.

Widely distributed on southern and western shores, northwards to Shetland, eastwards to Isle of Wight; recorded in Ireland from Kerry, Clare and Galway.

British Isles to at least Portugal; Mediterranean.

Probably perennial; gametangia recorded for April–July, cystocarps recorded for April and from July–January; tetrasporangia recorded throughout the year, with a peak from June–August. In Portugal, Ardré (1970) recorded spermatangia in June, cystocarps in March and October and tetrasporangia in March, April, June and October.

Because of uncertainities in identification, little information is as yet available on form variation in the British Isles. In the Mediterranean, the species appears to be very polymorphic and was treated as an aggregate by Boudouresque & Denizot (1975). More recently, however, Marcot & Boudouresque (1977) and Marcot-Coqueugniot (1980) have described some of the variants as new species, *P. rara-avis* and *P. hongii*.

P. harveyana may be more widely distributed than present records suggest, e.g. in the western Atlantic. It appears to be closely related to *P. inamoena* Pilger which is widely distributed in warm temperate and tropical seas.

Specimens of a *Peyssonnelia* dredged by Harvey and M'Calla from Bertraghboy Bay, Co. Galway and identified as *P. dubyi* (Harvey, 1846) have been seen in a number of herbaria. They were later reidentified by Batters (1896) as *P. rubra*, but Cotton (1912) disagreed with Batters after examining the type of *P. rubra*. The specimens are in fact referable to *P. harveyana*, although they are not the type of the species. Batters's identification was curious as he had recorded *P. harveyana* from Berwick in 1890.

The peculiarly-shaped carposporangia illustrated by Denizot (1968, Fig. 100) are probably aberrant; those figured by Ardré (1970, p. 7 fig. 1) appear to be developing from the connecting filament recalling those of *P. immersa*, q.v.

Peyssonnelia immersa Maggs & L. Irvine (1983), in press.

Holotype: BM. Isotypes: BM, GALW. Galway (Carraroe).

Crusts always adherent with lobed or entire margins, to 200 mm or more in extent, adjoining plants becoming confluent, up to 370 μm thick, bright red to brown or yellowish, drying darker, smooth with faint radial striae; rhizoids unicellular, thick-walled, 8 μm in diameter, to 25(40) μm long; basal filaments in parallel radial or rather sinuous rows, often irregular on uneven substrates, each cell 2–5 times longer than broad, cutting off a single cell from whole upper surface: these cells are boot-shaped in radial vertical section and give rise to compact branched filaments inclined towards the margin at an angle of more than 60° to the basal layer, becoming erect and about 8 μm in diameter above, growth zones visible in older specimens; calcification hypobasal, usually 10–15 μm thick.

Gametangial plants monoecious; spermatangial sori immersed, with a mucilaginous surface when mature, spermatangia colourless, borne in conspicuous terminal groups among pigmented filaments; carpogonia in the same sori, carpogonial and auxiliary cell branches borne laterally 6–8 cells below apex of little-modified, simple pigmented filaments, some cells elongated, uppermost cells broad, forming a roof to the sorus; gonimoblast of 4 or more rows of 3–5 carposporangia arising from connecting filament linking several auxiliary cells, carposporangia 20–25 μm, elevating the sorus surface somewhat when mature; tetrasporangial sori extensive, to at least 5 mm, little raised even when mature, tetrasporangia among simple pigmented sterile filaments 80–100 μm long composed of 5–7 cells, terminal with a conspicuous tetrasporangial supporting cell partly included within the wall, frequently borne on thickened filaments, 37–45×25–28 μm (including wall), with cruciately arranged spores, repeated generation of tetrasporangia common, accompanied by the production of additional wall layers in the sporangium.

Epilithic, on shells and epiphytic on maerl, in pools in lower littoral, sublittoral to 19 m in areas moderately sheltered from wave action.

Dorset, Argyll, Shetland Isles; Clare, Galway.

France (Roscoff).

Perennial; spermatangia recorded for April–November, carpogonia for January, May–August, October, cystocarps for June–November and tetrasporangia for January, March, April, June–November.

Thickness of the crust varies considerably; plants on shells are only 50–150 μm thick whereas those on rock grow up to 370 μm; this appears to be due to the greater age of plants on rock since they also show growth zones.

This species is to be expected elsewhere in the British Isles and surrounding areas. It could be confused with *Haematocelis rubens* J. Ag. from which it differs by having cruciate tetrasporangia and a calcified undersurface.

Palmariales

PALMARIALES
by
Linda M. Irvine
and
Michael D. Guiry*

PALMARIALES Guiry & Irvine

PALMARIALES Guiry & Irvine in Guiry (1978), p. 138.

Thalli variable in form, pseudoparenchymatous, the constituent filaments ranging from loose to compact; of multiaxial construction. Monoecious or dioecious; when dioecious spermatangial plant and tetrasporangial plant of similar organization; carpogonial plant (where known) very much smaller, carpogonia apparently occurring as single cells; tetrasporangial plant or tetrasporangia developing directly from fertilised carpogonium without a carposporophytic phase, and sometimes overgrowing carpogonial plant; the presence of a generative stalk cell beneath each tetrasporangium is a feature unique to this order.

The single family, Palmariaceae, shows affinities with certain families (especially Acrochaetiaceae) of the Nemaliales, an order which is a somewhat artificial grouping containing families which are extremely diverse in structure and reproduction (see 1(1) p. 75). Van der Meer (1981, 1981a) considered that it was more constructive to retain the Palmariaceae in a distinct order and that there was nothing to suggest that the diphasic life history was derived from a triphasic pattern. Ultrastructural studies of a range of species in these and other families (Pueschel, 1979, 1980; Pueschel & Cole, 1980, 1981, 1982 and 1982a) have provided characters which support Guiry's (1974a, 1978, Guiry & Irvine, 1981) taxonomic revisions and also suggest that revision of the Nemaliales is necessary.

PALMARIACEAE Guiry

PALMARIACEAE Guiry (1974a), p. 522.

Thallus solid or hollow, very variable in form; multiaxial, cortical cells in a single layer or in rows, medullary cells large, hyaline, loosely coherent, subisodiametric, in one or more layers. Gametangial plants monoecious or dioecious, spermatia borne in confluent sori over extensive areas of thallus, formed singly in contiguous spermatangia developed singly or in pairs on a spermatangial mother cell; carpogonia (where known) apparently occurring as single cells in young plant, tetrasporangia or tetrasporangial plant developing directly from fertilized carpogonium; carposporophyte lacking; tetrasporangia scattered or in sori, large, cruciate, with a generative stalk cell, formation accompanied by elongation and/or division of the vegetative cortical cells.

Three genera, *Halosacciocolax, Palmaria* and *Rhodophysema*, occur in the British Isles.

* Department of Botany, University College, Galway, Republic of Ireland

Rhodophysema has been transferred from the Peyssonneliaceae (Cryptonemiales) because a reinvestigation of the life history and reproduction of Californian material of *R. elegans* (Crouan frat. ex J. Ag.) Dixon in culture (DeCew, 1981; DeCew & West, 1982) has shown that this genus should be referred to the Palmariales. It is provisionally placed in the Palmariaceae, but there are a number of characters which suggest that it should be removed to a separate family. The affinities of the genus *Halosacciocolax* are uncertain (Pueschel & Cole, 1982a); it has been placed here pending further studies on the relationships between this family and the Acrochaetiaceae.

HALOSACCIOCOLAX

HALOSACCIOCOLAX S. Lund (1959), p. 192.

Type species: *H. kjellmanii* S. Lund (1959), p. 193.

Thallus growing on various members of the Palmariales, reputedly parasitic, consisting of branched endophytic filaments giving rise to a surface growth of radiating prostrate filaments which produce branched erect filaments.

Gametangial plants unknown in the British Isles; tetrasporangia terminal on erect filaments, cruciate, sterile filaments few or absent.

This genus has been placed in various families: see Lund (1959, Squamariaceae = Peyssonneliaceae), Parke & Dixon (1976, Rhodymeniaceae), Cabioch & Guiry (1976, Acrochaetiaceae), Pueschel & Cole (1982a, Palmariaceae). In a survey of the Acrochaetiaceae, Garbary (1979) considered the degree of morphological elaboration too great for placement in this family. It has been included provisionally in the Palmariaceae here pending further studies on the circumscriptions of and the relationships between these families.

For a comparison of this genus with *Neohalosacciocolax*, described for the Aleutian Is., see Lee & Kurogi (1978); the latter genus appears to be a typical member of the Palmariaceae.

One species in the British Isles:

Halosacciocolax kjellmanii S. Lund (1959), p. 193.

Lectotype: C (not seen). Greenland (Scoresby Sund).

Halosacciocolax lundii Edelstein (1972), p. 251.

Thallus on *Palmaria palmata* (L.) O. Kuntze, reputedly parasitic, cushion-like, up to 150 μm high and 3 mm in diameter, whitish when young, brownish when older, composed of irregularly branched endophytic filaments creeping between the outer cells of the host with occasional branches penetrating them, giving rise to a prostrate plate of radiating filaments which produce short branched erect filaments 10–12 μm in diameter, at first covered with a thin mucilaginous sheath which ruptures later.

Gametangia unknown in the British Isles; most erect filaments becoming fertile, protruding through ruptured sheath, tetrasporangia terminal, 30–38×20–25 μm, spores cruciately to irregularly arranged, secondary sporangia frequently developing from the same mother cell.

On old fronds of the host, especially near the base.
Anglesey, N. Devon, Cork, Waterford.

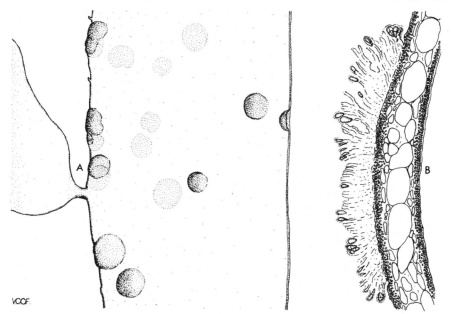

Fig. 20   *Halosacciocolax lundii*
A. Several plants on *Palmaria palmata* (May) × 8; B.V.S. tetrasporangial plant on host
(April) × 80.

Spitsbergen, Iceland, British Isles, France (Brittany); Greenland to Canada (Nova Scotia); USA (Washington).

Recorded in April and May in the British Isles and at other times of the year also elsewhere.

The young stages of this species, in which the cushions are much smaller and lighter in colour and the filaments are more delicate and penetrate the host tissue to a lesser degree, were described as *H. lundii* by Edelstein (1972); see Guiry (1975).

The parasitic nature of this species is suggested by the pale colour and the presence of filaments penetrating the host cells (see Guiry, 1975).

Spermatangia were reported in Greenland plants by Lund (1959). He described them as 'hyperparasitic' on the tetrasporangial plants, the spermatangia being borne on clavate mother cells terminating erect filaments (cf. Lee & Kurogi, 1978).

PALMARIA Stackhouse

PALMARIA Stackhouse (1801), p. xxxii.

Type species: *P. expansa* Stackhouse (1809), p. 69 (= *P. palmata* (Linnaeus) O. Kuntze (1891), p. 909; see Ruprecht (1850), p. 76).

Thallus consisting of a discoid base and an erect, stipitate frond with a flattened blade; blade branching dichotomous or palmate, marginal bladelike proliferations frequent; structure multiaxial with a pseudoparenchymatous medulla surrounded by cortical cells which increase in number rather than size as the frond matures. Gametangial plants dioecious; spermatangia in large superficial sori on large plants similar to tetrasporangial plants, mother cells produced in pairs from outer cortical cells; carpogonia apparently occurring as single cells in young plantlets only; tetrasporangial plant developing directly from fertilized carpogonium and overgrowing carpogonial plant, carposporophyte lacking; tetrasporangia interspersed with pigmented sterile filaments in large sori, cruciate, with a generative stalk cell.

One species in the British Isles:

Palmaria palmata (Linnaeus) O. Kuntze (1891), p. 909.

Lectotype: L 910.184.2889. Locality unknown.

Fucus palmatus Linnaeus (1753), p. 1162.
Palmaria expansa Stackhouse (1809), p. 69.
Palmaria lanceolata Stackhouse (1809), p. 69.
Rhodymenia palmata (Linnaeus) Greville (1830), p. xlix, 93 (as *Rhodomenia palmata*).

Thallus with a discoid holdfast and erect fronds, solitary or a few together, simple below or branching from the base, stipe inconspicuous, rarely to 5 mm long, blade gradually expanding above, dichotomously or palmately divided into broad segments, the total length to 500 mm (rarely to 1 m), width about 30–80 mm (rarely to 160 mm); blade sometimes simple, with marginal proliferations often dichotomous and large, resembling primary blade; colour purplish red, texture leathery-membranous, 150–250 μm thick, increasing to 350 μm in the tetrasporangial areas.

Structure multiaxial; cortex consisting of two layers of closely packed cells 8–10 μm in diameter and rather angular in surface view when young, increasing to about 10 cells thick in older or fertile regions, the cells being more or less similar in size and with frequent cell fusions; medulla composed of one or two layers of loosely arranged, almost isodiametric cells 120–150 μm in diameter.

Spermatangial sori indistinct, colourless or yellowish, scattered over most of frond, spermatangia elongate, about 10×5–6 μm when mature, secondary spermatangia frequent, one elongate spermatium liberated from each spermatangium. Carpogonial plants very small, narrow and stunted or encrusting, carpogonia becoming colourless, apparently occurring as single cells in young plants, tetrasporangial plant developing directly from the carpogonium after fertilization and overgrowing the carpogonial plant. Tetrasporangia in extensive, dark, cloud-like sori scattered over the whole frond; cortex becoming thicker, terminal cortical cells dividing to form a stalk cell and a tetrasporangial initial, the surrounding cells becoming modified to form curved pigmented sterile filaments; tetrasporangia 55–72×45–55 μm, spores cruciately arranged, about 30 μm when shed; stalk cell generative, producing a second mother cell and stalk cell in the cavity of the first.

Epilithic and epiphytic, especially on *Laminaria* stipes; littoral and sublittoral to a depth of 20 m in both sheltered and moderately exposed areas.

Generally distributed throughout the British Isles, but apparently absent from short stretches of coast in eastern England.

Arctic Russia to Portugal; Baltic. Arctic Canada to USA (New Jersey); USA (Alaska to California); Japan, Korea.

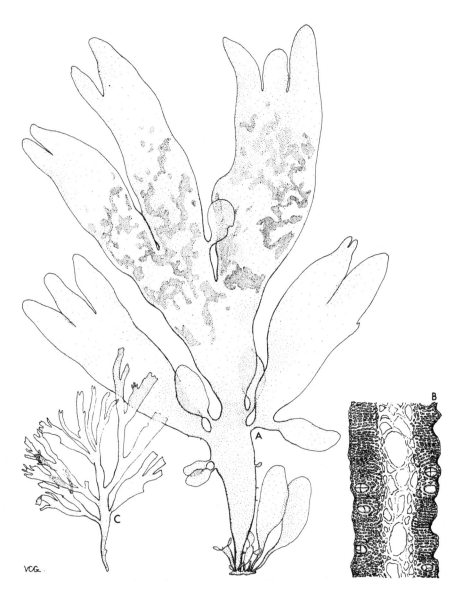

Fig. 21   *Palmaria palmata*
A. Habit of typical tetrasporangial plant (Mar.) × 1; B.T.S. tetrasporangial plant (Feb.)
× 80; C. Habit of plant sometimes referred to as var. *sarniensis* (July) × 1.

Perennial; young plants often cover the littoral early in the year, but those at higher levels rarely reach maturity; rapid growth of older plants begins in February/March (Austin, 1960) and proliferations, which are at first thin and pale, develop from the margins; plants capable of regenerating from fragments of a previous season's growth; spermatangia reported from (November) February–April (June), carpogonia known only in culture; tetrasporangia reported from November–April (rarely June–August).

Blade very variable in shape, probably due to the interaction of age and ecological factors and to its propensity for regeneration after damage; plants with simple blades have been called *Rhodymenia palmata* var. *simplex* (C.Ag.) Harv., whilst those with marginal prolifer-ations have been called *R. palmata* var. *marginifera* (Turn.) Harv. Sometimes the blade divisions are wedgeshaped and finely dissected above (*R. palmata* var. *sobolifera* (M.Vahl) Harv.) or the blade has numerous linear divisions throughout (*R. palmata* var. *sarniensis* (Roth) Grev.) This phenomenon seems to occur under fairly sheltered, silty conditions. Such plants are difficult to identify without examining the anatomical structure and the cortical cells in surface view, and have been confused with *Callophyllis cristata* (L. ex Turn.) Kütz. q.v. and *Gracilaria foliifera* (Forsk.) Børg. (see 1(1) p. 208).

Van der Meer & Todd (1980) gave the first description of carpogonial plants in this species and of the life history which they observed in culture. Carpogonia developed on sporelings a few days old grown from tetraspores released by plants from New Brunswick, Canada. They commented on the extreme sexual dimorphism shown and postulated that, in the field, the carpogonia are fertilized by spermatia from plants of a previous generation. See also van der Meer & Chen (1979) and van der Meer (1981, 1981a).

Fronds frequently bear algal epiphytes and endophytes and a number of marine fungi but more rarely the parasite *Halosacciocolax kjellmanii* Lund (Guiry, 1974, as *H. lundii* Edelstein) q.v.; galls are produced by nematodes (Barton, 1891), copepods (Harding, 1954) and bacteria (Chemin, 1927). The brown alga *Stictyosiphon griffithsianus* (Le Jol.) Holmes & Batt. is reported to occur only on this species (Mathias, 1935).

P. palmata is eaten or used as fodder for a variety of animals in many countries, being known as Dulse, Dillisk or Crannach in the British Isles. Percival (1979) reported that this species contains a water-soluble xylan but no floridean starch, (but see Pueschel, 1979); for a review of chemical constituents, see Morgan *et al.* (1980).

RHODOPHYSEMA Batters

RHODOPHYSEMA Batters (1900), p. 377.

Type species: *R. georgii* Batters (1900), p. 377.

Rhododermis J. Agardh (1851), p. XII, non *Rhododermis* Harvey (1844), p. 27.

Thallus thin and encrusting to spherical and cushionlike, closely adherent to substrate, rhizoids absent; basal layer monostromatic, cell fusions frequent, erect filaments little branched, the cells sometimes all remaining small or sometimes differentiating into a medulla of large colourless cells below.

Monoecious; spermatangia produced by single or paired mother cells terminal on erect filaments; carpogonia one-celled, terminal on erect filaments, dividing after fertilization to form a tetrasporangium initial and a generative stalk cell, carposporophyte lacking; tetra-sporangia in sori scattered over upper surface, cruciate, interspersed with sterile filaments.

A useful table showing features used in the discrimination of specific and infraspecific taxa was given by Tokida (1934) and amended later (Tokida, 1954) because he found that the size of the tetrasporangia is unusually variable. This is also shown by a comparison of the data given in recent papers by Cabioch (1975), Fletcher (1975, 1977), Ganesan & West (1975), Masuda & Ohta (1975) and South & Whittick (1976). These studies showed what appeared to be a simple recycling of the tetrasporangial phase but DeCew (1981) and DeCew & West (1982) found that, in Californian material of *R. elegans*, the tetrasporangia develop directly from fertilized carpogonia, previously interpreted as 'hair cells', on monoecious plants. Masuda & Ohta (1981) summarised earlier work and gave a useful comparison of and key to the seven described species of *Rhodophysema*.

KEY TO SPECIES

1 Epilithic, epizoic or epiphytic but not on *Zostera*; flat, without large-celled
   medulla... 2
   Epiphytic on *Zostera*; frequently forming subspherical cushions up to 1 mm
   with a large-celled medulla... *R. georgii*
2 Sterile filaments colourless, 4–5 cells long .... *R. elegans*
   Sterile filaments pigmented, 20 or more cells long .... *R. feldmannii*

Rhodophysema elegans (Crouan frat. ex J. Agardh) Dixon (1964), p. 70.

Holotype: LD 27725. France (Brest).

Rhododermis elegans Crouan frat. ex J. Agardh (1851), p. 505, nom illeg.
Rhododermis elegans forma *polystromatica* Batters (1890), p. 311.
Rhododermis polystromatica (Batters) Batters (1896), p. 389.
Rhodophysema elegans var. *polystromatica* (Batters) Dixon (1964), p. 71.
Peyssonnelia rupestris Crouan frat. (1867), p. 148.
Rhododermis parasitica Batters (1890), p. 312.

Crusts mainly smooth, often coalescent, to 45 mm or more in extent, and to 600 μm thick; rose red to dark red or purplish; erect filaments compact, 5–13 μm in diameter, cells 1–3 times as long, slightly tapering, little branched, cell fusions usually present, cells 4–5 μm in surface view, thallus with an outer mucilaginous layer up to 11 μm thick.

Spermatangia unknown in the British Isles; carpogonia formed at periphery of sori, tetrasporangial initials and mature tetrasporangia developing towards centre; sori visible with a ×10 lens as pale ruffled patches, giving the surface a watermarked appearance when dry; tetrasporangial primordia terminal, increasing to 55×30 μm after division, containing 4 cruciately arranged spores or, apparently, 2 uninucleate bispores, interspersed with colourless, curved, 4–6-celled sterile filaments 35–60(80)×6–8 μm.

Epilithic, epiphytic and epizoic; in sheltered intertidal pools and sublittoral to 9 m, found by Fletcher (1977) to be a common crust in sublittoral areas in southern England. It occurs on a wide variety of unusual substrates including glass, dogfish eggcases, and live crabs and molluscs.

Northwards to W. Ross, eastwards to Sussex; E. Scotland; widely distributed in Ireland; probably occurring throughout the British Isles.

Iceland, Faroes; Norway (Trondelag) to northwest France; W. Baltic; Mediterranean. Greenland, Canada (Newfoundland) to USA (Mass.); N. Pacific.

Probably perennial; tetrasporangia recorded throughout the year. Fletcher (1977) found that in culture the rate of growth varied with conditions, the greater part of the thallus remaining monostromatic under 'winter' conditions; the plants became fertile after 14 months. South & Whittick (1976) found that under their culture conditions of higher light intensity new plants became 20 cells thick in six months and three generations were produced in nine months.

The thickness of the thallus is very variable; the name *R. elegans* var. *polystromatica* (Batt.) Dixon has been applied to thalli 8–15 cells thick and *Peyssonnelia rupestris* Crouan frat. is also a synonym for this form (Denizot, 1968). *Rhododermis parasitica* Batters was applied to even thicker (12–30-celled) thalli growing on *Laminaria* stipes. Cabioch (1975) recorded from northern France *R. minus* Hollenberg & Abbott (1965), a species distinguished chiefly by its monostromatic habit. *R. elegans* frequently has extensive monostromatic regions and is obviously closely similar. The identity of *Cruoriopsis ensisae* Jao (1936) also needs to be investigated in this connexion.

Masuda & Ohta (1981a) discussed the discovery of distinct genetic races in *R. elegans* in Japan, based on differences in the number of cell layers in the crust, the dimensions of vegetative and reproductive cells and the life history pattern. They drew attention to the importance of orientation of sections (as in *Peyssonnelia*, q.v.). It remains to be seen whether the criteria used are taxonomically valid or a reflection of ontogenetic or life history states. It is possible that *Rhododermis elegans* var. *zostericola* Batters (1906) was based on non-medullate plants of *Rhodophysema georgii*, q.v.

Rhodophysema feldmannii Cabioch (1975), p. 109.

Holotype: PC. Isotype: BM. France (Roscoff).

Thallus consisting of a deep rose to dark red monostromatic crust less than 1 mm in diameter, adhering strongly to the substrate, rhizoids absent; cells in radial rows, 10–15 μm long and 5–7 μm broad, lateral fusions present.

Gametangial plants unknown, tetrasporangia in sori 150–200 μm in diameter, tetrasporangia 20–30 μm, terminal on 3- or more-celled filaments, with cruciately arranged spores;

Fig. 22   *Rhodophysema elegans*
A. Crust on stone with tetrasporangial sori appearing as ruffled patches (May) × 1; B. V.S. tetrasporangial sorus (Nov.) × 300.
Rhodophysema georgii
C. Several plants on *Zostera* leaf (June) × 8; D. Same showing tetrasporangia × 80.

intermixed with straight or slightly curved paraphyses up to 150 μm long, composed of 20 or more pigmented cells 7–8 μm in diameter.

Epizoic on the hydroid *Amphisbetia operculata* (L.) growing on stipes of *Laminaria hyperborea* (Gunn.) Fosl.; sublittoral.
Galway (Kilkieran Bay).
France (le Conquet and Roscoff).

Recorded for May in Ireland, February–March in France. Cabioch found that the tetraspores in culture produced crusts identical with the parent crust.
Data on seasonal growth and form variation insufficient for further comment.
This species recalls *Audouinella concrescens* (Drew) Dixon (see 1(1), p. 87) which is also epizoic and has closely similar dimensions. The so-called paraphyses of *R. feldmannii* are pigmented and are much longer than those of other species of *Rhodophysema*. Cabioch pointed out that the spore germination and crust development are different from that described by West (1970) for *A. concrescens*. Fletcher (1975, 1977), however, found two different spore germination patterns in *Rhodophysema* spp. A closely related species, *Audouinella spetsbergensis* (Kjellm.) Woelk. has also been recorded (as *Rhodochorton penicilliforme* (Kjellm.) Rosenv.) on *Amphisbetia operculata* in the Dinard–St Malo region by Lami (1940) and Miranda (1932), who discussed its relationship with the Cryptonemiales on the basis of spore germination and crust formation.

Rhodophysema georgii Batters (1900), p. 377.

Lectotype: BM. Scilly Isles.

Rhododermis van-heurckii Heydrich (1903), p. 243.

Thallus either crustose or cushionlike and then hemispherical to irregularly globose, up to 1 mm, dark purplish red, smooth, somewhat mucilaginous, occurring in groups, sometimes confluent; cushions with large medullary cells up to 200 μm in diameter, surrounded by little-branched cortical filaments with cells about 10 μm in diameter, tapering slightly to 6 μm in surface view.
Spermatangia unknown in the British Isles; carpogonia formed at periphery of sori with tetrasporangial initials and mature tetrasporangia developing towards centre; tetrasporangia in sori, terminal, 20–40×15–30 μm, spores cruciately arranged, interspersed with colourless, curved, 3–4-celled sterile filaments up to 50 μm long with cells 6–9×3–4 μm.

Epiphytic on *Zostera* leaves, especially along the margins; lower littoral and upper sublittoral. Recorded on other hosts outside the British Isles.
Recorded eastwards to the Isle of Wight, northwards to the Isle of Man; Channel Isles; Dublin, Clare, Galway and Mayo; probably occurring throughout the British Isles.
Norway (Trondelag) to northern Spain; western Baltic. Canada (Newfoundland) to USA (Long Is.), N.W. Pacific.

Probably perennial, the cushions recorded in summer. Fletcher (1975) found that the crusts became fertile in culture after 9–10 weeks, although they had not produced cushions.
Apparent intermediates between the crustose and cushion-like forms occur. The conditions under which the cushions with a large celled medulla develop are not known. Fletcher (pers. comm.) has been unable to find any distinguishing features between the crustose form

and *R. elegans* except that the germinating spores of the latter sometimes produce 'knot' filaments.

Masuda & Ohta (1975) reported this species growing on other red algae. Some of their plants produced spermatangia in sori without sterile filaments.

Rhodymeniales

RHODYMENIALES
by
Linda M. Irvine
and
Michael D. Guiry*

RHODYMENIALES Schmitz

RHODYMENIALES Schmitz in Engler (1892), p. 19.

Thalli crustose and/or erect and frondose, constituent filaments compact, pseudoparenchymatous; of multiaxial construction, hollow or solid.

Procarpic, carpogonium formed from the apical cell of an accessory 3- or 4-celled branch, auxiliary cell branches 2-celled, auxiliary cell cut off before fertilization from a daughter cell of the carpogonial branch supporting cell but not fully differentiating until after fertilization, carposporophyte development following transfer of the zygote nucleus to an auxiliary cell or cells either directly or via a connecting cell, gonimoblast development always outwards, cystocarp always protuberant with a well-developed pericarp, carposporangia liberating a single carpospore; tetrasporangia intercalary or terminal in cortical filaments, cruciate or tetrahedral; gametangial plant and tetrasporangial plant of similar organization.

Opinions on the distinctness of this order vary considerably, some workers regarding it as sharply defined while others stress the similarity with certain families of the Gigartinales such as the Gracilariaceae; see 1(1) p. 61; Guiry & Irvine, 1981. Representatives of two families occur in marine situations in the British Isles:

Champiaceae
Rhodymeniaceae.

CHAMPIACEAE Kützing

CHAMPIACEAE Kützing (1843), p. 439 [as Champieae].

Champioideae (Kützing) Kylin (1931), p. 28 [as Champieae].

Thallus erect or reflexed, at least partially hollow, terete, compressed or flat, much branched; multiaxial, with a compact cortex and central cavity lined by longitudinal parallel or anastomosing medullary filaments bearing secretory cells, cavity divided either by monostromatic septa or by multilayered plugs throughout or at branch bases; procarpic, carpogonial branches 3- or 4-celled, with one or two 2-celled auxiliary cell branches, fusion cell present, carpogonia produced terminally or from all gonimoblast cells, cystocarps developing externally, inner wall of distended pericarp sometimes stretched to form a reticulum of delicate filaments (*tela arachnoidea*), pore prominent or absent; tetra(poly) sporangia in cortex, intercalary or terminal, sometimes in invaginated sori.

Four genera are represented in the British Isles: *Champia, Chylocladia, Gastroclonium*

* Department of Botany University College, Galway, Republic of Ireland

and *Lomentaria*. The anatomical features, such as the nature of the medullary filaments and the presence of septa, used for distinguishing these genera (see Bliding (1928), Kylin (1956) and Lee (1978)) can be seen by using a stain such as iodine in KI, preferably after cutting the thallus longitudinally. Some species with polysporangia have been placed in a separate genus, *Coelosira* Hollenberg, but there appears to be no justification for this (see Chang & Xia, 1978, who transferred them to the genus *Gastroclonium*). For further comparisons within and between genera, see Reedman & Womersley (1976) and Lee (1978).

CHAMPIA Desvaux

CHAMPIA Desvaux (1809), p. 245.

Type species: *C. lumbricalis* (Linnaeus) Desvaux (1809), p. 246.

Mertensia Roth (1806), p. 318, nom. illeg., non *Mertensia* Roth (1797), p. 34 nec *Mertensia* Willdenow (1804), p. 165.
Corinaldia Trevisan (1843), p. 334.

Thallus with a discoid holdfast and erect or prostrate fronds, hollow throughout, terete or compressed, slightly constricted into regular segments by septa, branching usually alternate, sometimes irregular, opposite or whorled, not deciduous; structure multiaxial, cortex with a tubular inner layer of large cells covered by an incomplete layer of small cells; medulla consisting of single-layered pseudoparenchymatous septa separating each segment and a number of regularly arranged, discrete longitudinal filaments lining cortex, bearing secretory cells adaxially and uniting with septa.

Gametangial plants dioecious; spermatangial sori superficial, spermatangia subterminal on elongate mother cells formed from cortical cells; procarpic, carpogonial branches in cortex, 4-celled, each with a single 2-celled auxiliary cell branch, gonimoblast developing outwards from a fusion cell, only terminal cells forming carposporangia, enclosed within distended cortical pericarp, *tela arachnoidea* well developed, cystocarps scattered, external, with a well-defined pore; tetrasporangia intercalary in cortical filaments, irregularly scattered in younger parts of thallus, tetrahedral, polysporangia unknown.

One species in the British Isles:

Champia parvula (C. Agardh) Harvey (1853), p. 76.

Lectotype: LD (Herb. Alg. Agardh. 26022). Spain (Cadiz).

Chondria parvula C. Agardh (1824), p. 207.

Fronds arising in tufts from a small discoid base, usually matted and entangled, mucilaginous but firm, to 100 mm long, terete or slightly compressed, 1–2 mm broad, pinkish to dull red or greenish brown; segments shorter than broad in younger parts, elongating to 1–2 diameters in older parts, septa usually visible externally, causing slight constrictions; branching variable, usually alternate, sometimes opposite or secund, whorled only when dense, branches often patent, arising at, or slightly above, the septa at intervals of a few segments, rarely from adjacent segments, narrowed below but not markedly constricted at point of insertion, apices rounded.

Structure multiaxial; medullary filaments parallel, connecting with the septa, 500–700×10–15 μm, secretory cells *c*. 15 μm in diameter, cortical cells axially elongated, covered by an incomplete layer of cells 10–30 μm in surface view which may be absent in the younger parts.

Gametangial plants dioecious; spermatangia 2–3 μm, developing serially, borne superficially in sori forming girdles around one or more segments which thus appear swollen and pale; cystocarps subspherical, to about 1 mm, with a well-defined pore and well-developed *tela arachnoidea*; carposporangia conical, 50–120 μm long in a cluster to 450 μm broad, enveloping filaments absent; tetrasporangia numerous in the cortex in younger segments, protruding into the medulla when mature, 55–120 μm, spores tetrahedrally arranged.

Epiphytic in pools in the upper sublittoral.

This species appears to be more restricted in distribution than previous reports indicated.

Fig. 23   *Champia parvula*
A. Habit (July) × 1⅓; B. Branch with cystocarps (Sep.) × 6 (note pores).
Chylocladia verticillata
C. Habit (June) × 1⅓; D. Branch with cystocarps (Aug.) × 6 (pores absent).

Its presence has so far been confirmed only from Cornwall eastwards to Sussex; Channel Isles; in Ireland from Cork northwards to Mayo. Reports from elsewhere have been found to be based on misidentification of other algae, particularly *Chylocladia verticillata* (Lightf.) Bliding.

British Isles to Portugal; Mediterranean; widely reported from warmer parts of the Atlantic, Indian and Pacific Oceans, but species limits are at present ill-defined.

Annual, most abundant in summer, but a few plants found throughout the year; spermatangia recorded for July and October, cystocarps for August–September and tetrasporangia from July–September. Steele & Thursby (1981) reported that all stages were easily maintained in laboratory culture and completed their life cycle in 8–10 weeks.

The variation in appearance is mainly due to the irregular nature of the branching.

This species is easily confused with small plants of *Chylocladia verticillata* (Lightf.) Bliding. In *Champia parvula* the branches are rarely whorled, narrow at the base but not markedly constricted and never arise from more than 3 adjacent segments; the segments are half to twice as long as broad and the cystocarps have a prominent pore. In *Chylocladia verticillata* the branches are usually whorled, markedly constricted and paler at their insertion and usually arise from every segment except near the apices; most segments are at least twice as long as broad and the cystocarps have no pore.

CHYLOCLADIA Greville in W. J. Hooker nom. cons.

CHYLOCLADIA Greville in W. J. Hooker (1833), p. 256, 297.

Type species: *C. kaliformis* (Goodenough & Woodward) Greville in W. J. Hooker (1833), p. 297 (= *C. verticillata* (Lightfoot) Bliding (1928), p. 69).

Kaliformis Stackhouse (1809), p. 56.
Kaliformia Stackhouse (1816), p. IX.
Gastridium Lyngbye (1819), p. 68, nom. illeg., non *Gastridium* Beauvois (1812), p. 21.

Thallus with a small attachment disc and erect or arching fronds, hollow throughout, but divided into segments by septa, terete or slightly compressed, constricted at septa, branching variable, often whorled, not deciduous; structure multiaxial, resulting in a tubular inner cortex of large cells covered by an incomplete layer of smaller cells and lined inside by discrete parallel longitudinal medullary filaments bearing secretory cells adaxially and uniting with monostromatic septa between the segments.

Gametangial plants dioecious; spermatangia in superficial sori, terminal on mother cells derived from cortical cells; procarpic, carpogonial branches in cortex, 4-celled, each with two 2-celled auxiliary cell branches, cystocarps protruding externally, carposporangia large, wedge-shaped, arising directly from large central fusion cell, enclosed within cortical pericarp without a pore, *tela arachnoidea* absent; tetrasporangia scattered in younger parts, intercalary in cortical filaments, tetrahedral, polysporangia unknown.

Lomentaria patens Kütz. (as *Chylocladia kaliformis* var. *patens* (Kütz.) Harv.) and *C. squarrosa* (Kütz.) Le Jol. have been reported for the British Isles; they are included here as doubtful synonyms of *C. verticillata* (Lightf.) Bliding. Both were originally described from the Mediterranean (see Ercegović, 1956) together with a number of other related species all of which are in need of further investigation throughout their ranges before the names can be correctly applied.

One species in the British Isles:

Chylocladia verticillata (Lightfoot) Bliding (1928), p. 69.

Lectotype: BM-K. Argyll (Jura).

Fucus verticillatus Lightfoot (1777), p. 692.
Fucus kaliformis Goodenough & Woodward (1797), p. 206.
Gigartina pygmaea Lamouroux (1813), p. 137 (reprint p. 49).
Lomentaria pygmaea (Lamouroux) Gaillon (1828), p. 367 (reprint p. 19).
Chylocladia kaliformis (Goodenough & Woodward) Greville in W. J. Hooker (1833), p. 297.
?*Lomentaria squarrosa* Kützing (1843), p. 440.
?*Lomentaria patens* Kützing (1843), p. 440.
?*Chylocladia kaliformis* var. *patens* (Kützing) Harvey (1851), pl. 358B, fig. 1, 2.
?*Chylocladia kaliformis* var. *squarrosa* (Kützing) Harvey (1851), pl. 358B, fig. 3.
?*Chylocladia squarrosa* (Kützing) Le Jolis (1863), p. 142.

Frond arising from a small attachment disc, erect, often pyramidal, usually with a percurrent axis, mucilaginous but firm, to 300(550) mm long, terete, hollow, 1–5 mm broad, pink, red, purple, yellowish in well-illuminated situations; segments at least as long as broad in younger parts, much longer below, up to 5(10) diameters, constrictions at septa giving a distinctly beaded appearance to younger parts; branching usually repeatedly whorled, occasionally distichous, opposite, branches erect or patent, occasionally recurved, arising above the septa consecutively from every segment, absent only in the younger parts, markedly constricted and usually pale at the point of insertion, apices acute, adventitious branchlets occurring in luxuriant specimens.

Structure multiaxial, medullary filaments parallel, uniting with the septa, secretory cells c. 15 μm, cortex 2-layered, outer layer often very incomplete, cells 10–30 μm in surface view.

Gametangial plants dioecious; spermatangia 2–3 μm, in pale superficial sori at first forming girdles in the grooves between the segments, later covering the segments entirely; cystocarps external, subspherical, about 500 μm, without a pore or *tela arachnoidea*, carposporangia wedge-shaped, about 150 μm long, arranged like segments of an orange in a cluster about 450 μm broad around a large fusion cell, carpospores 55–80 μm on discharge; tetrasporangia enlarging after division, subspherical, 45–200 μm, intercalary in the cortex of younger segments, spores tetrahedrally arranged, 55–80 μm on discharge.

Epilithic and epiphytic, littoral especially in pools and sublittoral to at least 12 m; tolerant of high light intensity.

Widely distributed on southern and western shores, eastwards to Sussex, northwards to Shetland, but records from the east coast few; in Ireland eastwards to Wexford; Dublin; northwards to Donegal; Antrim, Down.

Norway (N. Trondelag) to Morocco; Sweden, Denmark; Mediterranean; Canary Isles.

Plants probably mainly annual; sporelings typically appearing December–February and degeneration of thallus obvious in mature specimens by July/August; spermatangia recorded for July–September, cystocarps from April–October and tetrasporangia from May–September. In culture, we found plants grew rapidly and completed their life history (carpospore to carpospore) in about 12 weeks at 20°C.

Much variation occurs in this species; it is not known to what extent this is pheno- or genotypically determined, but preliminary culture studies suggest that temperature, light intensity and day length are important factors. Culture results indicate that 2–3 branches are produced at the constrictions at first, with further branches developing secondarily, the arrangement often appearing whorled; in very old thalli, adventitious branches can be produced between the whorls. Plants with short, horizontal branches in closely set whorls have been called *Chylocladia squarrosa* (Kütz.) Le Jol. (= *C. kaliformis* var. *squarrosa* (Kütz.) Harv.); such plants may bear a superficial resemblance to *Lomentaria articulata* (Huds.) Lyngb., q.v. Plants with long, widely separated, attenuated, often distichous branches have been called *Chylocladia kaliformis* var. *patens* (Kütz.) Harv.; these plants bear a superficial resemblance to forms of *Lomentaria clavellosa* (Turn.) Gaill., q.v.

Small plants of *Chylocladia verticillata* are difficult to distinguish from *Champia parvula* (C. Ag.) Harv., q.v. Buffham (1888) recorded tetrasporangia and cystocarps in the same plant from Sidmouth, August 1886. Similar plants collected from Finavarra, Clare (July, 1979) appeared to be primarily cystocarpic plants with tetrasporangial branches possibly arising from carpospores that had germinated on the plants.

GASTROCLONIUM Kützing, nom. cons.

GASTROCLONIUM Kützing (1843), p. 441.

Type species: *G. ovale* (Hudson) Kützing (1843), p. 441 (= *G. ovatum* (Hudson) Papenfuss (1944), p. 344).

Sedoidea Stackhouse (1809), p. 57, 83.
Dasyphylla Stackhouse (1816), p. IX.
Coelosira Hollenberg (1940), p. 871.

Thallus with a branched or discoid holdfast and fronds consisting of an erect solid terete stipe and hollow, erect or reflexed, terete or compressed deciduous branches which are sometimes short and vesicle-like, sometimes longer and segmented by septa, branching dichotomous or lateral and variable; structure multiaxial, hollow branches with a tubular inner cortex of large cells covered by an incomplete layer of smaller cells and lined inside by discrete, parallel longitudinal medullary filaments bearing secretory cells adaxially and uniting with monostromatic septa.

Gametangial plants dioecious, reproductive organs developing only on the hollow branches; spermatangia in superficial sori, terminal on elongated mother cells borne on branched filaments derived from cortical cells; procarpic, carpogonial branches 4-celled, with two 2-celled auxiliary cell branches, gonimoblast developing outwards, carposporangia large, developing directly from a large fusion cell, cystocarps protruding externally, cortex forming a thick pericarp without a pore, *tela arachnoidea* absent; tetrahedral tetrasporangia or polysporangia intercalary in cortical filaments, in sori in younger parts.

KEY TO SPECIES

Solid stipe more than 10 mm long, hollow branches usually vesicle-like, rarely segmented, not reflexed; bearing normal tetrasporangia .... *G. ovatum*
Solid stipe less than 5 mm long, hollow branches always segmented, many reflexed; bearing polysporangia with 8 spores .... *G. reflexum*

Gastroclonium ovatum (Hudson) Papenfuss (1944), p. 344.

Lectotype: original description (Hudson, 1762, see Irvine & Dixon, 1982) in the absence of material. Yorkshire.

Fucus ovatus Hudson (1762), p. 468.
Chylocladia ovata (Hudson) Batters (1902), p. 73.
Fucus ovalis Hudson (1778), p. 573.
Chylocladia ovalis (Hudson) Greville in W. J. Hooker (1833), p. 297.

Thallus with a branched holdfast and erect fronds to 150 mm long, consisting of an irregularly dichotomously branched terete solid main axis bearing terete or slightly compressed hollow branches above, to 7(10) mm long, and 1–2 mm broad, which are vesicle-like and usually spherical at first becoming elongate later, sometimes divided into 2–4 segments by septa and appearing beaded; firm to subcartilaginous; dark brownish or purplish red, occasionally greenish, vesicles translucent.

Structure multiaxial; vesicles with a 2-layered cortex, outer layer often incomplete, cells 10–30 μm in surface view, medullary filaments parallel, 15 μm in diameter, secretory cells

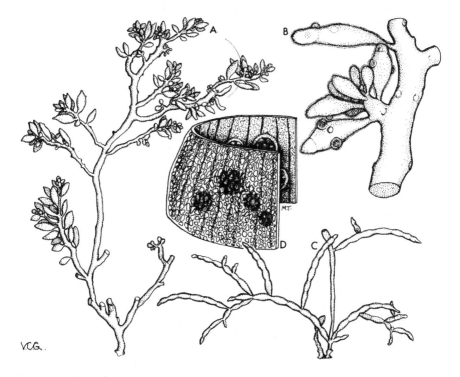

VCG.

Fig. 24   *Gastroclonium ovatum*
A. Habit of mature plant with both old and young vesicles (June) × 1⅓; B. Branch with cystocarps (July) × 6
Gastroclonium reflexum
C. Habit (July) × 1⅓; D. Part of branch cut away to show polysporangia and medullary filaments lining cavity (Aug.) × 80.

15 μm; solid axis with a pseudoparenchymatous large-celled medulla and cortex of 1–2 layers of smaller, radially elongate cells.

Gametangial plants dioecious; spermatangia in pale superficial sori on vesicles, 4–5 μm; gonimoblast with a large fusion cell bearing sessile carposporangia arranged like segments of an orange, 120–140×80 μm, wedge-shaped, cystocarps external, subspherical, about 600 μm, without a pore or *tela arachnoidea*, spores shed by local disintegration of pericarp; tetrasporangia 65–85 μm, intercalary in cortex of vesicles, spores tetrahedrally arranged.

Epilithic and epiphytic, particularly on *Corallina*; littoral and upper sublittoral, most commonly in pools in areas with some exposure to wave action.

Generally distributed throughout the British Isles, northwards to Orkney, but rarely recorded on eastern coasts.

British Isles to Mauritania; Canary Isles; records from the Mediterranean, Colombia and Brazil may represent separate entities.

Recorded throughout the year, possibly perennial with deciduous vesicles; spermatangia recorded from August–September, cystocarps from May–September and tetrasporangia from March–December with a peak in summer.

There is great variation in vesicle size, but the factors controlling this are not known; in the most luxuriant specimens they are segmented like the branches of *G. reflexum, G. clavatum* (Roth) Ardiss., a Mediterranean species, and *Chylocladia verticillata* (Lightf.) Bliding. Such plants have been referred to *Chylocladia ovata* var. *subarticulata* (Turn.) Batt. and have been recorded mainly from April–July; they seem to be plants in which the vesicles have continued elongating since the previous summer. During the summer, some plants can be difficult to identify because the old vesicles have been shed and new vesicles are just beginning to develop; others have a mixture of both, as described by Lightfoot (1777, as *Fucus vermicularis* S. G. Gmelin).

Buffham (1888) reported polysporangia containing 16 spores in plants from Teignmouth, August 1881, collected by W. H. Gilburt, but the material does not seem to have been preserved.

Gastroclonium reflexum (Chauvin) Kützing (1849), p. 866.

Lectotype: CN (herb. Lenormand). France (Port-en-Bessin).

Lomentaria reflexa Chauvin (1831), fasc. 6, no. 143 (see Sayre, 1969).
Lomentaria pygmaea sensu Duby (1830), p. 950 pro parte, non *Gigartina pygmaea* Lamouroux (1813), p. 137 (reprint p. 49) nec *Lomentaria pygmaea* (Lamouroux) Gaillon (1828), p. 367.
Chylocladia reflexa (Chauvin) Zanardini (1843), p. 50.

Thallus with an irregular discoid to branched holdfast and fronds consisting of a short solid stipe up to 3 mm long bearing a tuft of erect and reflexed hollow, terete branches to 60 mm long and 2 mm broad, constricted by septa, giving a distinctly beaded appearance, mucilaginous but firm, dark red, purple or greenish; segments at least as long as broad in younger parts, much longer later, up to 5 diameters; secondary branching dichotomous, distichous or secund, branches markedly constricted and usually paler at point of insertion, apices acute; reflexed branches often stolon-like and producing new attachment discs.

Structure multiaxial, branches with medullary filaments parallel, uniting with septa, secretory cells 12–15 μm, cortex 2-layered, outer layer often incomplete, cells 12–15 μm in surface view; stipe with a cortex of small pigmented cells and a medulla of large colourless cells.

Gametangial plants dioecious; spermatangia 3×1 μm, in pale superficial sori on the younger segments; cystocarps about 500 μm, external, clustered on erect branches, subspherical, without a definite pore or *tela arachnoidea*, carposporangia wedge-shaped, 180–200×25–100 μm, spores released by local distintegration of pericarp; polysporangia 60–120 μm, spherical, numerous in younger segments of erect branches, usually containing 8 regularly arranged spores.

Epilithic and epiphytic, especially on *Corallina*; littoral and upper sublittoral; iridescent specimens often in sandy, well-lit pools.

Recorded for Cornwall, Devon, Sussex; Pembroke, Glamorgan; Wexford, Galway; Channel Isles.

British Isles to Portugal; Canary Isles; eastern Mediterranean.

Plants have been recorded between April and December but little is known of their seasonal behaviour. Reproduction seems to be confined to the summer months; spermatangia recorded for August, cystocarps from August–September and polysporangia from June–September. The reflexed branches frequently form new attachment discs and these probably provide a means of vegetative propagation and perennation.

Data on form variation too inadequate for comment.

Guiry, Cullinane & Whelan (1979) found that polysporangia were confined to the upper 5 mm of erect branches in Irish plants. Although Buffham (1888) reported 16 spores in the polysporangia, 8 spores have been found consistently in British Isles material examined. Ardré (1970) reportéd and illustrated sporangia containing 8 or 12 spores in Portuguese plants. Miranda (1931) studied the development of 8-spored polysporangia in Spanish plants and considered that the arrangement of the nuclei suggested that meiosis had occurred. Normal tetrasporangia have been found on specimens from the eastern Mediterranean in BM.

This species is often confused with young plants of *Chylocladia verticillata* (Lightf.) Bliding which are frequent in pools in February–March. *G. reflexum* has a solid stipe to 3 mm long which is sometimes branched; in *C. verticillata* the solid stipe does not extend much beyond the hapteron and is less than 1 mm in extent. *G. reflexum* is purplish red whilst *C. verticillata* is brownish red when young.

G. reflexum is closely related to *G. clavatum* (Roth) Ardiss., a Mediterranean species also iridescent and with a tuft of branches at the top of the stipe, but without the reflexed habit.

LOMENTARIA Lyngbye

LOMENTARIA Lyngbye (1819), p. 101.

Type species: *L. articulata* (Hudson) Lyngbye (1819), p. 101.

Chondrothamnion Kützing (1843), p. 438.
Chondrosiphon Kützing (1843), p. 438.
Hooperia J. Agardh (1896), p. 89.

Thallus usually erect, sometimes arching or partly prostrate, terete or compressed, hollow throughout, sometimes constricted into segments, branching variable, not deciduous; structure multiaxial, cortex tubular, composed of 3–6 layers of cells lined inside with a network of medullary cells bearing secretory cells adaxially and compacted into a plug at constrictions and/or branch bases, monostromatic septa absent.

Gametangial plants dioecious; spermatangia in superficial sori, terminal on mother cells derived from cortical cells; procarpic, carpogonial branches 3-celled, with one or two (one non-functional) 2-celled auxiliary cell branches, gonimoblast developing outwards, most cells forming carposporangia, *tela arachnoidea* poorly developed, cystocarps scattered, protruding externally, with a prominent pore; tetrasporangia terminal on cortical filaments, in sori in depressions formed by invagination of cortex, tetrahedral, polysporangia unknown.

In *L. articulata* the multilayered plugs constrict the medullary cavity at regular intervals throughout the frond. In *L. clavellosa* and *L. orcadensis*, however, they occur only at the bases of the branches, and the frond has an unconstricted appearance. Guiry (unpubl.) has suggested the removal of species lacking a regularly constricted frond to a separate genus, for which the name *Chondrothamnion* Kütz, would be available.

KEY TO SPECIES

1   Branches constricted at intervals; apical branching often dichotomous .... *L. articulata*
    Branches not constricted at intervals; apical branching not dichotomous 2
2   Fronds up to 400 mm long, terete or compressed; holdfast discoid .... *L. clavellosa*
    Fronds up to 80 mm long, flat; holdfast usually stoloniferous................... *L. orcadensis*

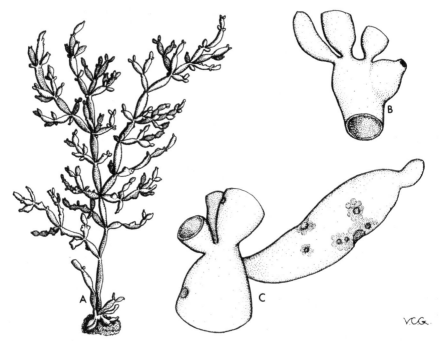

Fig. 25   *Lomentaria articulata*
           A. Habit of plant with cystocarps (April) × 1⅓; B. Branch with cystocarp (April) × 16
           (note pore); C. Branch with tetrasporangia lining depressions in cortex (Sep.) × 35
           (Anatomy and tetrasporangia as in Fig. 26.

Lomentaria articulata (Hudson) Lyngbye (1819), p. 101.

Lectotype: original description (Hudson, 1762) in the absence of material (see Irvine & Dixon, 1982). Cornwall.

Ulva articulata Hudson (1762), p. 476.

Thallus consisting of a small discoid holdfast and one or more erect, or occasionally arched, fronds up to 100 mm long; fronds terete or compressed, conspicuously constricted into segments, the segments hollow, ovate or clavate and up to 4(10) mm long and 1(1·5) mm in diameter; dark brownish to bright red, bleaching to pinkish orange, translucent, shiny and turgid; branching variable, dichotomous, alternate or opposite, sometimes whorled, some-times more or less distichous, branches borne at the constrictions.

Structure multiaxial; medullary filaments forming a network lining the cortex in the segments, compacted at each constriction into a 3–5 layered plug of large rounded cells, monostromatic septa absent, secretory cells *c.* 15 μm; cortex of 3–6 layers of cells, cells 6–10 μm in surface view.

Gametangial plants dioecious; spermatangial sori pale, covering the surface of younger segments; carposporangia in an irregular mass, carpospores 45–60×10–20 μm, cystocarps external, hemispherical to conical, 500–600 μm in diameter, with a thick cortical pericarp and prominent pore; tetrasporangial sori in the younger segments, in depressions in the cortex, tetrasporangia enlarging to about 60 μm in diameter, spores tetrahedrally arranged, 39–48 μm on discharge.

Epilithic and epiphytic, in and out of pools, particularly in shady places in the littoral and sublittoral to at least 18 m.

Generally distributed throughout the British Isles.

Faroes; Norway (Nordland) to Cameroun; Canary Isles; Mediterranean, Black Sea.

Perennial, prostrate branches probably capable of vegetative propagation; spermatangia reported for August, cystocarps for January, April–October and tetrasporangia throughout the year, with a peak in summer.

This is a distinctive species which is usually easily recognised; there is comparatively little morphological variation, except in the length of the segments and the degree of compression of the fronds. Exceptionally large, luxuriant plants have been found in the sublittoral, growing on *Laminaria* stipes together with normal sized plants, but no explanation for this can be offered at present.

Newton's (1931) report of iridescence in deep water has not been confirmed.

Chylocladia verticillata (Lightf.) Bliding may be confused with this species. The former has barrel-shaped segments, cystocarps without a pore and tetrasporangia not in cortical depressions; *L. articulata* has spindle-shaped segments frequently paired at the apices like rabbit ears, cystocarps with a prominent necked pore and tetrasporangia in cortical depress-ions. *Catenella caespitosa* (With.) L. Irvine in Parke & Dixon may also be confused with this species (see 1(1), p. 191).

Lomentaria clavellosa (Turner) Gaillon (1828), p. 367.

Lectotype: BM-K (see Turner, 1801, pl. 9). Norfolk.

Fucus clavellosus Turner (1801), p. 133.

Thallus with a small attachment disc and erect fronds up to 400 mm long and often pyramidal in outline, not segmented, hollow, main axis usually narrowed at the base, sometimes naked below, up to 4 mm in diameter, sometimes markedly compressed with the branching distichous and up to 4 times pinnate, sometimes terete with radial branches inserted irregularly, luxuriant specimens clothed with adventitious branchlets; plants soft to subcartilaginous, dark brownish or purplish red to pink; branches more or less tapering at base and apex.

Structure multiaxial; medullary filaments 12–18 μm in diameter, forming a network, closely compacted at branch bases into a 3–5-layered plug of large rounded cells, monostromatic septa absent, secretory cells c. 10 μm; cortex with an almost complete outer layer of slightly elongated cells 4–6 μm in surface view.

Gametangial plants dioecious; spermatangia 4–5 μm, cut off from elongated mother cells

Fig. 26   *Lomentaria clavellosa*
A. Habit of 'sheltered water' form (Aug.) × 1⅓; B. Habit of 'rough water' form (Feb.) × 1⅓; C. Branch with cystocarps (July) × 6; D. Part of branch showing tetrasporangia lining depressions in cortex and medullary filaments lining cavity (Sep.) × 80
Lomentaria orcadensis
E. Habit (Aug.) × 1⅓.

in pale sori in the younger parts; cystocarps external, conical, 375–550 μm, with a thick cortical pericarp and a prominent pore, carposporangia up to 60 μm; tetrasporangia in sori developing near the apices of the branches in depressions in the cortex, sporangia 45–60 μm in diameter when mature, spores tetrahedrally arranged, 30–32 μm when shed.

Epilithic and epiphytic; lower littoral in shady places in and out of pools and sublittoral to at least 24 m; in areas moderately to extremely exposed to wave action.

Generally distributed throughout the British Isles.

Iceland, Norway (Nordland) to Morocco; Faroes, Denmark, Sweden, Helgoland; Mediterranean, Black Sea; USA (New Hampshire–Connecticut).

Recorded throughout the year, individuals probably annual; preliminary observations suggest that most plants begin growth early in the year, become fertile in the summer and degenerate after the spores are shed, occasionally regenerating from the basal disc the following year; cystocarps recorded for January and April–November, tetrasporangia for January and March–September. Spermatangia have not been recorded for the British Isles and were reported as rare in Sweden by Kylin (1923).

The appearance of plants varies with the extent and regularity of the branching, degree of flattening and texture of the frond (Boillot, 1961). Plants are much stiffer in areas exposed to wave action and in extreme cases can be mistaken for a species of *Gelidium*. Flattened, distichous forms have been called *Chylocladia clavellosa* var. *sedifolia* (Turn.) Grev. ex Harv. in W. J. Hook. (= *Lomentaria sedifolia* (Turn.) Strømf. ex Fosl.) and *L. clavellosa* var. *pyramidalis* Thur. ex Le Jolis.

Small distichous plants are difficult to distinguish from *L. orcadensis*. In such plants the branches at the apex are about 500 μm apart and about 500 μm broad. In *L. orcadensis* the branches are 2–4 mm apart and about 1 mm broad. Soft delicate forms of *L. clavellosa* resemble *Gloiosiphonia capillaris* (Huds.) Carm. in Berk., q.v., but this latter species has internal cystocarps and tetrasporangia confined to an encrusting phase. Sterile specimens can be identified by a comparison of the anatomy, especially of the apex which in *Gloiosiphonia* does not show the discrete medullary filaments typical of the Champiaceae.

Lomentaria orcadensis (Harvey) Collins ex W. R. Taylor (1937a), p. 309.

Holotype: BM. Orkney (Skaill).

Chrysymenia orcadensis Harvey (1849), p. 100.
Chrysymenia rosea var. *orcadensis* (Harvey) Harvey (1850), pl. 301; 358A.
Chylocladia rosea (Harvey) Harvey (1853), p. 186.
Lomentaria rosea (Harvey) Thuret ex Farlow (1881), p. 155.

Thallus with prostrate and erect fronds, prostrate fronds filiform, terete, branched, giving rise to erect fronds sometimes filiform, usually flattened and triangular in outline; not mucilaginous, to 30(80) mm long, hollow; main axis to 5(10) mm broad, tapering at base and apex, pinkish to dark red; branching opposite or alternate, distichous, with first and second order lateral branches resembling main axis, patent, often naked below.

Structure multiaxial, medullary filaments 3–5 μm in diameter forming a network, compacted into a plug of large rounded cells at branch bases, monostromatic septa absent, secretory cells 10–12 μm, cortex with an incomplete outer layer of cells 4–8 μm in surface view.

Gametangial plants unknown; tetrasporangia in sori in depressions in the cortex, 33–55×25–40 μm, spores tetrahedrally arranged, c. 26 μm in diameter when shed.

Epilithic, epiphytic, usually on *Laminaria* stipes, and epizoic on sponges; from the upper sublittoral in deep shady pools to at least 20 m.

Widely distributed throughout the British Isles; literature records comparatively few as the species is easily overlooked.

Iceland, Faroes; Norway (N. Møre) to Portugal; Heligoland, Denmark; Canada (Nova Scotia) to USA (N. Carolina).

Plants probably perennial; prostrate fronds probably capable of vegetative propagation; tetrasporangia recorded from February–September, individuals rarely sterile; tetraspores in culture produced tetrasporangial plants similar in morphology to the parent plant (Foran & Guiry, 1983).

The erect fronds of this species have a distinctive appearance, with little variation. The prostrate system is sometimes poorly developed in epilithic specimens.

Svedelius (1937) reported apomeiosis in this species; Magne (1964), however, found a chromosome number of n = 10 or 20 in plants from different localities in Brittany. Gametangial plants reported by Crouan frat. (1867), Segawa (1936) and Lodge (1948) are probably based on other species of *Lomentaria*.

Small specimens of *L. clavellosa* (q.v.) have been confused with this species.

RHODYMENIACEAE Harvey nom. cons. prop.

RHODYMENIACEAE Harvey (1849), p. 75, 120.

Thallus erect, encrusting or parasitic, erect fronds terete, compressed or flattened, solid in British Isles representatives; branching monopodial or apparently sympodial, dichotomous or proliferous; structure multiaxial, medulla compact and pseudoparenchymatous, cortex compact with cells in single layers or in dichotomously branched filaments; procarpic, with a 2- or 3-celled auxiliary cell branch formed before fertilization and a 3- or 4-celled carpogonial branch, connecting cell occasionally formed, gonimoblast developing outwards, most cells transformed into carposporangia, cystocarps large, protuberant, formed on surface of thallus, with a pore, *tela arachnoidea* present or absent; tetrasporangia formed in cortex between sterile filaments, terminal or intercalary, cruciate or rarely tetrahedral.

Three genera occur in the British Isles: *Cordylecladia*, *Rhodymenia* and *Rhodymeniocolax*. the structure and known reproduction of *Cordylecladia* accords almost exactly with that of *Rhodymenia*; however, *Cordylecladia* is distinguished morphologically by its almost terete erect fronds arising in groups from an encrusting base and swollen, apical sori. *Rhodymeniocolax* is a genus of apparently parasitic plants which has only recently been found in the British Isles and northern France; it was previously unknown outside the Pacific. Another parasitic genus, *Halosacciocolax*, was included in the Rhodymeniaceae by Parke & Dixon (1976) because when *H. lundii* (now = *H. kjellmanii* Lund) was first recorded for the British Isles (Guiry, 1974) it was thought to be related to its host, *Rhodymenia palmata* (L.) Grev. (now *Palmaria palmata* (L.) O. Kuntze). It has been transferred to the Palmariaceae provisionally (see Pueschel & Cole, 1982), although Cabioch & Guiry (1976) considered it may have affinities with the Acrochaetiaceae.

CORDYLECLADIA J. Agardh

CORDYLECLADIA J. Agardh (1852), p. 704.

Type species: *C. erecta* (Greville) J. Agardh (1852), p. 704.

Thallus with an expanded base giving rise to several terete to slightly compressed erect fronds, branching sparse, dichotomous to irregular; structure multiaxial, medulla pseudo-parenchymatous with axially elongated cells, cortex of several rows of smaller cells decreasing in size outwards.

Gametangial plants dioecious; spermatangia in superficial sori at swollen apices, formed in pairs on mother cells derived from cortical cells; details of carpogonial and auxiliary cell branches unknown, gonimoblast often lobed, developing outwards, most cells becoming carposporangia, enveloping filaments absent, cystocarps large, elevating cortex and protruding externally, with a pore; tetrasporangia intercalary in cortex, in sori at or near swollen apices, cruciate.

Cordylecladia erecta (Greville) J. Agardh (1852), p. 704.

Lectotype: E. Devon (Sidmouth). Paratypes: BM-K, LD. Devon (Torbay). See Jones (1962).

Sphaerococcus ? erectus Greville (1828), pl. 357.
Gracilaria erecta (Greville) Greville (1830), p. 124.

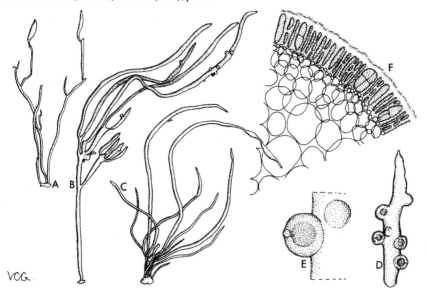

VCG.

Fig. 27   *Cordylecladia erecta*
A. Habit of plant with tetrasporangial sori (April) × 1; B. Habit of plant with cystocarps (July) × 1; C. Habit of plant with spermatangial sori (April) × 1; D. Branch with cystocarps (July) × 6; E. Enlargement of cystocarp to show pore × 16; F. T.S. showing tetrasporangia (April) × 80.

Thallus with an expanded discoid base to 30 mm broad and 60 μm thick giving rise to several erect, simple or branched terete fronds to 110 mm long, bright to dark red or brownish, never purple, translucent, rigid and brittle; branches few, irregular, at a narrow angle, little tapering, up to 100 mm long and 500–800(1500) μm broad.

Structure multiaxial; medulla pseudoparenchymatous, cells large, up to 25–60 μm in diameter, 50–200 μm long axially, cortex of radial rows of smaller cells decreasing in size outwards, axially elongated in surface view, 5–9×5–7·5 μm.

Gametangial plants dioecious; spermatangia 4–5×2–3 μm, in whitish pod-like structures at the apices; gonimoblast with an elongated basal cell, most cells becoming carposporangia 18–22×15–20 μm, grouped in successively developing gonimolobes, without enveloping filaments, cystocarps protruding, spherical, up to 800 μm in diameter, with a small pore, discharged carpospores 20–25 μm; tetrasporangia in the cortex, in sori confined to swollen pod-like branch apices, 30–50×15–30 μm, spores cruciately arranged, 21–22 μm on discharge.

Epilithic, lower littoral and sublittoral to 32 m; on sandy rocks, in pools and channels, apparently tolerating complete burial by sand.

Northwards to Orkney, eastwards to Kent; Channel Isles; widespread in Ireland except on the east coast between Armagh and Wexford.

British Isles to northern Spain.

Plants recorded throughout the year, probably perennial, the tips frequently dying back in the summer after fruiting, and regenerating; individuals become fertile in early autumn; spermatangia recorded for September, December and January, cystocarps recorded for September, November–April, tetrasporangia for November–April. There is probably also extensive vegetative propagation from the crustose base.

Apart from degree of branching, little form variation has been reported for this species.

Carpospores in culture produced plants similar in morphology to the parent plant. This species has been placed in the genus *Gracilaria* (see 1(1), p. 205) but was shown by Feldmann (1967) to belong to the Rhodymeniaceae; she also showed that Mediterranean reports were based on misidentifications of *Gracilaria* spp. In the British Isles specimens of *C. erecta* are sometimes confused with *Gracilaria verrucosa* (Huds.) Papenf. (see 1(1), p. 210) especially as the latter also grows in sandy places.

RHODYMENIA Greville nom. cons.

Rhodymenia Greville (1830), p. XLVIII, 84 (as *Rhodomenia*).

Type species: *R. palmetta* (Lamouroux) Greville (1830), p. XLVIII, 84 (= *R. pseudopalmata* (Lamouroux) P. C. Silva (1952), p. 265.

Dendrymenia Skottsberg (1923), p. 16.

Thallus with erect or prostrate, usually stipitate fronds, arising from a basal disc or stolons, blades flattened, simple or divided dichotomously, palmately or irregularly; sometimes with marginal or apical proliferations; structure multiaxial, medulla compact, pseudoparenchymatous, with large axially elongated cells, cortex of radial filaments of 2–5 smaller cells.

Gametangial plants dioecious; spermatangia in small subapical sori or large irregular patches scattered over blade, produced superficially from outer cortical cells; carpogonial branches 3–4-celled, supporting cell also bearing 2-celled auxiliary cell branch, gonimoblast developing outwards often with 2–3 lobes, almost all cells forming carposporangia, enveloping filaments absent; cystocarps hemispherical, large and protruding, with a pore, formed apically or scattered; tetrasporangia in subapical sori or scattered throughout blade, intercalary in unmodified cortex, cruciate.

The British Isles species of *Rhodymenia* were revised by Guiry (1977); he listed two species *R. delicatula* and *R. pseudopalmata*, the latter having two varieties: *pseudopalmata* and *ellisiae*. He discussed the differences between the varieties and pointed out that some specimens are apparently intermediate. It is, however, current taxonomic practice to award similar entities elsewhere species status (Dawson, 1941, Sparling, 1957), as it is generally recognised that there is a worldwide problem in delimiting species in this genus. In the present work, therefore, these two taxa are treated as species, *R. pseudopalmata* and *R. holmesii* respectively. Specimens of a prostrate species have occasionally been found in southern England (Kent, Devon). More recently, prostrate specimens with tetrasporangia in special outgrowths have been found at 20–28 m in the sublittoral in Donegal (Maggs & Guiry, 1982) and in similar locations in Galway, Cork and Pembrokeshire. Reports of a Mediterranean species, *R. ardissonei* Feldm. (= *R. corallicola* Ardiss. pro parte), from the British Isles were based on misidentifications (see Dixon 1964a). The specimens were probably of *R. holmesii*, which is closely related to *R. ardissonei*. *Palmaria palmata* (Linnaeus) O. Kuntze (q.v.) was for many years placed in *Rhodymenia* (as *Rhodymenia palmata* (L.) Grev.) but Guiry (1974a, 1978) has shown that it does not belong to the Rhodymeniales.

Species of *Rhodymenia* have often been confused with other genera, especially *Phyllophora* (see 1(1), p. 221). Guiry (1977) gave a useful comparison of the cortical cells in surface view of three species of *Rhodymenia* and nine species of other genera with a similar morphology.

KEY TO SPECIES

1 Mature plant less than 25 mm long and 80 μm thick 10 mm from apex .... *R. delicatula*
   Mature plant up to 80(120) mm long and 200 μm thick 10 mm from apex 2
2 Plants without extensive stoloniferous growth; blades over 5 mm broad 10 mm
      from apex; cystocarps (sub)apical on blade............................... *R. pseudopalmata*
   Plants with extensive stoloniferous growth; blades less than 5 mm broad 10 mm
      from apex; cystocarps usually on lower half of blade .... *R. holmesii*

Rhodymenia delicatula P. J. L. Dangeard (1949), p. 172.

Lectotype: Laboratoire de Botanique, Talence, Bordeaux (see Dangeard, 1949, fig. 10N; Guiry, 1977, fig. 42). Morocco (Agadir).

Thallus consisting of a peg-like attachment disc and erect fronds with a slender cylindrical stipe expanding into a simple or once or twice dichotomous blade; fronds to 15(22) mm long, blades to 3 mm broad, (30)55–85(100) μm thick, often incurved when mature, light rose-red in colour; stolons sometimes present.

Structure multiaxial; medulla of 2–3 layers of cells *c*. 20–60 μm in diameter, cortical cells widely spaced, about 18–14×7–11 μm in surface view.

Spermatangia unknown; cystocarps at the base of the blade, small, hemispherical, up to 500 μm, gonimoblast lobed, enveloping filaments absent, surrounded by a thick cortical pericarp with a small pore, carposporangia 16–21 μm, carpospores 12–15 μm when shed;

tetrasporangia in small cortical sori at the base of the blade, intercalary, 24–35×16–23 μm, spores cruciately arranged.

Epilithic; in deep, shady pools well below the water surface and on open rock and maerl in the lower littoral and sublittoral to a depth of at least 20 m; usually in sandy or silty areas.

Recorded for Isle of Wight, S. and N. Devon, Glamorgan, Anglesey, Orkney and Galway; probably widely distributed throughout the British Isles but easily overlooked because of its small size.

British Isles to Morocco.

Young fronds developing in winter and spring, sometimes from stolons and also regenerating from eroded stipes or blade apices; spermatangia unknown, cystocarps and tetrasporangia recorded from December–May, July, with most of the growth of the frond occurring after the formation of the reproductive organs.

There is some variation in width and degree of subdivision of the blades which may be ecologically determined.

Gall-like protuberances have been reported to occur on the surface of the frond; these are whitish red, sometimes branched and have the appearance of a collapsed balloon; they may be vegetative reproductive structures. *Rhodymenia phylloïdes* L'Hardy-Halos (1970), described from northern France, is very similar to this species but is reported to have tetrasporangia and cystocarps in a subapical position.

Rhodymenia delicatula superficially resembles *Phyllophora traillii* Holmes ex Batt. and young plants of *Schottera nicaeensis* (Duby) Guiry & Hollenb. (see 1(1), p. 226–231) but may be distinguished from these species by the more gradual expansion from stipe to blade, the light rose-red colour of the blades and the absence of marginal proliferations. For confirmation and determination of very young and sterile specimens microscopical examination is necessary; the surface cortical cells of *R. delicatula* are of irregular size and are relatively large whilst those of *S. nicaeensis* and *P. traillii* are more regular in size and smaller (cf. Guiry, 1977, figs. 2–13).

Rhodymenia holmesii Ardissone (1893), p. 682.

Holotype: BM. Sussex (Hastings) (see Guiry, 1977 fig. 31).

Halymenia palmetta var. *ellisiae* Duby (1830), p. 943.
Rhodymenia palmetta var. *ellisiae* (Duby) Bornet (1892), p. 286 (reprint p. 126) (as *elisiae*).
Rhodymenia pseudopalmata var. *ellisiae* (Duby) Guiry in Guiry & Hollenberg (1975), p. 149.
Rhodymenia nicaeensis sensu Holmes (1883), p. 289 non *R. nicaeensis* (Lamouroux ex Duby) Montagne
  (1846), p. 68 (= *Schottera nicaeensis* (Lamouroux ex Duby) Guiry & Hollenberg (1975), p. 153).

Thallus with an extensive, branched, stoloniferous holdfast and erect, flattened fronds, blade often fan-shaped, arising from a short stipe; fronds to 80 mm long, 5 mm broad and 100–150 μm thick, to 5 times regularly dichotomous, pinkish to purplish red, thin, flaccid, often spirally twisted and incurved, apices sometimes elongated into filiform proliferations to 20 mm long, marginal proliferations from base of plants common, often reattaching to the substrate.

Structure multiaxial, medulla pseudoparenchymatous with 2–6 layers of axially elongated cells 40–55 μm in diameter and 100–140 μm long, cortex of 2–3 layers of axially elongated cells 9–14×5–7 μm in surface view, subcortical cells visible, regularly arranged.

Gametangial plants dioecious, structure of spermatangia unknown; cystocarps usually present only at the bases of larger plants, 700–800 μm in diameter with a pore, gonimoblast

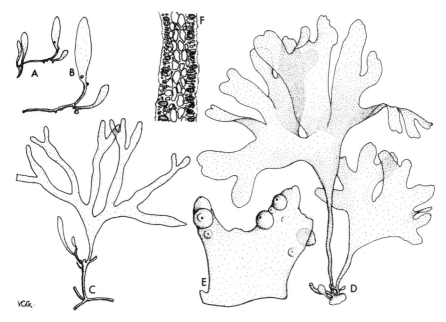

Fig. 28   *Rhodymenia delicatula*
A. Habit of cystocarpic plant (July) × 1; B. Same × 2
Rhodymenia holmesii
C. Habit (July) × 1
Rhodymenia pseudopalmata
D. Habit (July) × 1; E. Part of blade with cystocarps (July) × 6; F. T.S. showing tetrasporangia (June) × 80.

with 2–3 lobes, *tela arachnoidea* absent, spore mass to 500 μm leaving a gap between pericarpial wall and spores; tetrasporangia in small subapical sori and reputedly also in apical proliferations, 23–25(30)×14–20 μm, spores cruciately arranged.

Epilithic, stolons ramifying in soft rock and sponges, tolerant of sand cover and some exposure to wave action; lower littoral and sublittoral to at least 25 m.

Northwards to Anglesey and eastwards to Kent, Norfolk (drift); in Ireland northwards to Donegal, eastwards to Wexford.

British Isles to northern Spain and possibly Morocco.

Young fronds appear in February–March, arising from the stolons and reach full size about July; spermatangia recorded for November, cystocarps and tetrasporangia for April, June–September (see Guiry, 1977).

Filiform apical proliferations probably only occur in old plants after the apices have become eroded.

R. coespitosella L'Hardy-Halos (1976), described from Brittany, is very similar to *R. holmesii* but the fronds are deep purple-red, the stolons arise from a disc, the cystocarps are (sub)apical and the cortical cells are smaller in median parts of the blade (see L'Hardy-Halos, 1976 figs 18–23). Specimens answering this description have not been found in the British Isles.

Rhodymenia pseudopalmata (Lamouroux) P. C. Silva (1952), p. 265.

Lectotype: CN (see Guiry, 1977 fig. 1). Spain (La Coruña).

Fucus pseudopalmatus Lamouroux (1805), p. 29.
Delesseria palmetta Lamouroux (1813), p. 125 (reprint p. 37).
Rhodymenia palmetta (Lamouroux) Greville (1830), p. XLVIII, 88.

Thallus with several erect fronds arising from a discoid base up to 5 mm broad, each consisting of a simple or branched terete stipe gradually expanding into a fanshaped blade, branches from stipe sometimes behaving as stolons, stipe up to 10(40) mm long, blade up to 30(60) mm long with dichotomous, palmate or irregular divisions, pinkish or brownish red, cartilaginous, not mucilaginous, 100–200(250) μm thick, axils rounded, apices tapering, rounded or spathulate, especially when fertile, in old plants occasionally elongated into filiform appendages.

Structure multiaxial, medulla pseudoparenchymatous with large axially elongated cells up to 50 μm in diameter and 100–150 μm long, cortex of 2 or 3 layers of axially elongated cells 9–14×8–12 μm in surface view; subcortical cells, when visible, not appearing regularly arranged.

Gametangial plants dioecious; spermatangia in pale subapical sori 2–5(8) mm in diameter, formed in pairs on cortical mother cells, 5–11×4·5–9 μm; carpogonial branches 3-celled, cystocarps large, drying black, subapical, hemispherical, up to 1 mm, with a pore, spore mass 500 μm with a gap between wall and spores, enveloping filaments absent, carposporangia rounded-hexagonal (5)7–22(32)×(5)7–14(22) μm; tetrasporangia in swollen subapical sori which appear dark red, 25–30×15–22 μm, spores cruciately arranged, 20–29 μm after discharge.

Epilithic and epiphytic, often on the stipes of *Laminaria hyperborea* (Gunn.) Fosl. in shady pools and crevices in the lower littoral and sublittoral to at least 17 m.

Generally distributed on southern and western shores, northwards to Argyll and eastwards to Kent; in Ireland northwards to Mayo and eastwards to Wexford; Antrim.

British Isles to at least Morocco; Azores; Canary Isles; Mediterranean and South African records doubtful. Records from elsewhere apply to other species of *Rhodymenia*.

Stipe and fronds perennial; young plants appearing in February/March. Spermatangia recorded for June–September; cystocarps produced from June onwards, sometimes persisting until January or even March; tetrasporangia recorded for July to November, and January.

Fronds are very variable in the degree and angle of branching and shape of the blade divisions. Plants growing epiphytically have a fan-shaped blade arising abruptly from a long stipe and are said by divers to look very different from epilithic plants although no anatomical differences have been seen.

This species is frequently confused with *Phyllophora* spp., *Schottera nicaeensis* (Lamour. ex Duby) Guiry & Hollenb. and *Stenogramme interrupta* (C.Ag.) Mont. ex Harv. (see 1(1) pp. 221, 231, 233). When examined in surface view, all these phyllophoracean algae are seen to have more regular and generally smaller cortical cells.

The tetraspore size quoted by Boney (1975) seems abnormally small; the measurements

could possibly have been based on a member of the Phyllophoraceae (see 1(1) p. 211).
This species has been reported in the eastern Atlantic as far south as Angola (Lawson, John & Price, 1975) but African plants have not been fully studied on a comparative basis.

RHODYMENIOCOLAX Setchell

RHODYMENIOCOLAX Setchell (1923a), p. 394.

Type species: *R. botryoidea* Setchell (1923a), p. 394.

Thallus inconspicuous, base consisting of filaments (?parasitic) penetrating the host; erect fronds with short branches, pinkish, consistency slightly softer than host; structure multi-axial, medulla pseudoparenchymatous, surrounded by a cortical layer one to several cells in thickness.

Spermatangial plants unknown; carpogonial branches 4-celled in a procarp with a 2-celled auxiliary cell branch; cystocarps as in the genus *Rhodymenia*; tetrasporangia scattered over most of thallus surface, cruciate.

One species in the British Isles:

Rhodymeniocolax sp.

Thallus up to 5 mm in diameter, composed of a cluster of short, often densely and irregularly divided terete to slightly flattened branches up to 0·75 mm in diameter and 2 mm long arising from a cushion-like or obconical base which penetrates deeply into the thallus of *Rhodymenia* spp; plants pink to pinkish-white, of a firm consistency but not as firm as the host.

Structure multiaxial, consisting of a medulla of rounded to rounded-elongate cells 15–30×15–50 μm and a cortex of 1–5 layers of rounded to rounded-elongate cells 4–8×4–15 μm in surface view.

Cystocarps formed terminally or subterminally on elongated, terete to slightly flattened fronds, forming a distinct protuberance on the surface of these; only immature cystocarps known in British material; tetrasporangia and spermatangia unknown in the British Isles.

On *Rhodymenia* spp. from the upper sublittoral to 20 m.
Pembroke, Glamorgan, N. Devon, Cornwall.
France (Brittany: L'Hardy-Halos, pers. comm.).

Plants have been found from August–October most commonly on host plants growing on *Laminaria* stipes.

The plants are similar in structure and reproduction to *Rhodymeniocolax botryoidea* Setchell, the type and only previously known species, from the eastern Pacific. They differ from this species in that the thallus is slightly larger (up to 5 mm), is irregularly dichotomously branched with terete fronds, and the young plants generally have a stellate appearance, although older plants are more irregular.

There is considerable variation in shape in the few specimens available; some plants form irregular tumourous growths spreading over the thallus surface from the initial point of infection whilst others form discrete stellate thalli from a single attachment point.

Specimens have only recently been found in the British Isles and insufficient information is available to assess their taxonomic status. Plants were probably not collected previously because of their small size, inconspicuous nature and seeming occurrence only in the sublittoral. Other dichotomously branched irregular protuberances are occasionally formed on the surface of the thallus of *Rhodymenia* species but these growths are generally the same colour as the plant itself and do not bear reproductive organs; such protuberances may reproduce the thallus vegetatively.

GLOSSARY

Terms are included in this glossary either when these are not in general botanical use or where several different meanings are possible and we wish to clarify our particular usage. This list can be regarded as provisional and will be updated at the end of the Rhodophyta volume.

ACUMINATE   Tapering gradually to a point.

ADAXIAL   The side, usually the uppermost, of a lateral adjacent to the axis.

ADVENTITIOUS   Arising in an irregular manner or from an abnormal position.

APPRESSED   Pressed close.

AUXILIARY CELL   A cell which becomes joined to the carpogonium either by direct fusion or as a result of the growth of a connecting filament. In most cases gonimoblast filaments arise from the auxiliary cell after this junction.

BASIONYM   Name-bringing or epithet-bringing synonym.

BILOCULAR   With two cavities.

BISPORANGIUM   A sporangium in which two spores are formed.

BISPORE   One of the two spores formed in a bisporangium.

CAESPITOSE   Tufted.

CARPOGONIAL BRANCH   A filament terminated by the carpogonium.

CARPOGONIUM   A female sexual cell consisting of a basal portion, the contents of which function as a gamete, and an elongated apical receptive portion, the trichogyne.

CARPOSPORANGIUM   A sporangium produced by a carposporophyte and containing one spore (see carpotetrasporangium)

CARPOSPORE   A spore formed in a carposporangium.

CARPOSPOROPHYTE   A morphological phase consisting of growths originating from a carpogonium.

CARPOTETRASPORANGIUM   A sporangium containing four spores produced by a carposporophyte.

CARPOTETRASPORE   One of the four spores formed in a carpotetrasporangium.

CLAVATE   Club-shaped.

CONNECTING FILAMENT   Cell(s) by which connection is established between a carpogonium and an auxiliary cell. In a few cases gonimoblast development is initiated from the connecting filament, e.g. *Polyides* (see Kylin, 1956, p. 168; Maggs & Irvine, 1983).

CORIACEOUS   Leathery.

CORTEX   A term loosely used for the peripheral region of a thallus. See 1(1) p. 11.

CORTICATING FILAMENT   A filament forming a superficial investment.

CORYMBOSE   A flat-topped branch system.

CRUCIATE   Applied to the arrangement of spores in a tetrasporangium resulting from a transverse division of the initial, followed by a division of each of the products in a plane or planes at right angles to the first division. See 1(1) pp. 21, 22.

CRUCIFORM   In the form of cross.

CUNEATE   Wedge-shaped.

CYSTOCARP   A term used for various structures representing the whole carposporophyte or a part, with or without a pericarp.

DENDROID   Tree-like, as a result of lower branches falling away.

DETERMINATE AXIS   An axis of limited growth.

DIOECIOUS   With male and female gametangia on separate thalli.

DISTICHOUS   Arranged in two opposite rows.

DIVARICATE   Spreading widely.

ENDOPHYTIC   Living within a plant.

ENDOZOIC   Living within an animal.

ENVELOPING FILAMENTS Filaments surrounding a gonimoblast but produced by the female gametangial thallus.

EPILITHIC Living on rock or loose stones.

EPIPHYTIC Living on a plant, but attached to the surface only.

EPIZOIC Living on an animal, but attached to the surface only.

FILAMENT A branched or unbranched row of cells joined end to end.

FILIFORM Thread-like.

FLACCID Limp.

FOLIOSE With a broad, flat blade.

FROND That part of the thallus other than the attachment structure.

FURCATE Forked.

FUSION CELL A product of cell fusion, usually applied to the fusion of the carpogonium and/or the auxiliary cell with adjacent cells.

GAMETANGIUM A cell producing one or more gametes.

GAMETE A sexual cell capable of uniting with another sexual cell.

GAMETOPHYTE A morphological phase which bears gametangia.

GONIMOBLAST Tissue which develops from the carpogonium or an auxiliary cell and which ultimately produces carposporangia or carpotetrasporangia.

HAPTERON, pl. HAPTERA A specialized multicellular attachment structure.

HOLOTYPE The one specimen or other element used by the author or designated by him as the nomenclatural type.

HYPOGYNOUS CELL In the carpogonial branch, the cell next to the carpogonium.

INDETERMINATE AXIS An axis of potentially unlimited growth.

INTERCALARY Occurring in any position in a thallus or filament other than at apex or base.

ISOTYPE Any duplicate (part of a single gathering made by a collector at one time) of the holotype.

LACUNOSE With many holes.

LANCEOLATE Narrow and tapering at both ends.

LECTOTYPE A specimen or other element selected from the original material to serve as a nomenclatural type when no holotype was designated at the time of publication or as long as it is missing.

LIGULATE Strap-shaped.

LITTORAL Applied to that portion of the shore which is alternately exposed to the air and wetted either by the tide or by splash and spray. See 1(1) pp. 51–53.

LUBRICOUS Slippery.

MAERL Loose-lying corallines, either living or dead, usually aggregated into sublittoral masses.

MEDULLA A term loosely used for the internal region of a thallus; it may be pseudoparenchymatous or distinctly filamentous. See 1(1) p. 11.

MIDLITTORAL Applied to the middle portion of the littoral, q.v.

MONILIFORM Bead-like.

MONOECIOUS With male and female gametangia on the same thallus.

MONOSPORANGIUM A sporangium in which a single spore is formed.

MONOSPORE A spore formed in a monosporangium.

MONOSTROMATIC Composed of a single layer of cells.

MULTIAXIAL Of a thallus containing several axial filaments.

PAPILLA A short nipple-like outgrowth.

PARASPORE A spore formed in a parasporangium.

PARASPORANGIUM A sporangium of irregular and variable shape, producing a variable number of spores, usually more than 4; not homologous with a tetrasporangium.

PARENCHYMA A compact tissue formed by cell division in all planes.

PARIETAL Lying along the wall, peripheral to the cell.

PATENT Spreading widely.

PEDICEL A stalk of a reproductive organ.

PELTATE More or less circular and attached by the centre of the lower surface.

PERICARP That part of the cystocarp produced by the female gametangial thallus and forming a covering to the developing gonimoblast. It may be a massive compact structure, or adherent filaments, or loose filaments referred to as enveloping filaments.

PERICENTRAL CELL A cell developing as a lateral primordium from an axial cell.

POLYSIPHONOUS Consisting of axial cells each surrounded by a ring of cells of the same length.

POLYSPORE A spore formed in a polysporangium.

POLYSPORANGIUM A sporangium homologous with a tetrasporangium in which more than four spores are formed.

PROCARPIC With the carpogonium and auxiliary cell in the same specialized filament system.

PSEUDOPARENCHYMA A tissue formed by the aggregation of branched or unbranched filaments and having the appearance of parenchyma.

PYRENOID An organelle occurring within or adjacent to a chloroplast; often associated with reserve food accumulation.

PYRIFORM Pear-shaped.

RHIZINE A specialized thickwalled rhizoid.

RHIZOID A unicellular or multicellular filament formed secondarily, either externally (for attachment) or internally.

SECRETORY CELL An almost colourless cell with highly refractive contents usually occupying a definite position in the thallus. A variety of physiological processes are now known to be associated with these cells and in some cases they appear not to have a secretory function.

SECUND Arranged on one side only.

SESSILE Without a stalk.

SINUATE Of a plane structure with a wavy margin.

SORUS An aggregation of reproductive structures.

SPERMATANGIUM A male gametangium in which a single non-flagellated spermatium is formed.

SPERMATIUM Gamete formed in a spermatangium.

SPATHULATE Ovate with an attenuated base.

STIPE The lowermost stalklike part of an erect frond.

SUBLITTORAL Applied to that portion of the shore which is either totally immersed or only uncovered by the receding tide infrequently and then for very short periods. See 1(1) pp. 51–53.

SUPPORTING CELL A bearing cell; usually applied to the cell bearing one or more carpogonial branches.

SYMPODIAL A mode of development in which the apparent main axis is not developed by continuous apical growth, but is made up of the basal parts of successive lateral filaments or axes.

SYNTYPE Any one of two or more specimens cited by the author when no holotype was designated; or any one of two or more specimens simultaneously designated as types.

SYNONYM One of two or more names for the same taxon.

TAXON A taxonomic category of any rank.

TERETE Circular in transverse section.

TETRAHEDRAL Applied to the arrangement of spores in a tetrasporangium resulting from simultaneous division of a spherical initial into 4 equal spores with the apex of each pointing towards the centre of the tetrasporangium. See 1(1) pp. 21–22.

TETRASPORANGIUM A sporangium in which four spores are formed.

TETRASPORE One of the four spores formed in a tetrasporangium.

TETRASPOROPHYTE A morphological phase bearing tetrasporangia.

TRABECULAE Filaments traversing a cavity.

TRICHOGYNE The receptive apical prolongation of a carpogonium.

UNIAXIAL Of a thallus containing a single axial filament.

UNILOCULAR With one cavity.

UNISERIATE With cells arranged in a single row; a filament.

ZONATE Applied to the arrangement of spores in a tetrasporangium resulting from 3 parallel divisions of the initial. See 1(1) pp. 21–22.

REFERENCES FOR
CRYPTONEMIALES, PALMARIALES AND RHODYMENIALES

ABBOTT, I. A. 1967.   Studies in the foliose red algae of the Pacific Coast, I. Cryptonemiaceae. *J. Phycol.* **3:** 139–149.

ABBOTT, I. A. 1968.   Studies on some foliose red algae of the Pacific Coast, III. Dumontiaceae, Weeksiaceae, Kallymeniaceae. *J. Phycol.* **4:** 180–198.

ABBOTT, I. A. 1979.   Taxonomy and nomenclature of the type species of *Dumontia* Lamouroux (Rhodophyta). *Taxon* **28:** 563–566.

AGARDH, C. A. 1817.   *Synopsis Algarum Scandinaviae.*   Lundae.

AGARDH, C. A. 1822.   *Species Algarum* **1**(2).   Lundae.

AGARDH, C. A. 1822a.   *Icones Algarum Ineditae.*   Lundae.

AGARDH, C. A. 1824.   *Systema Algarum.*   Lundae.

AGARDH, J. G. 1841.   In historiam algarum symbolae.   *Linnaea* **15:** 1–50.

AGARDH, J. G. 1842.   *Algae Maris Mediterranei et Adriatici.*   Parisiis.

AGARDH, J. G. 1848.   Nya alger från Mexico.   *Ofvers. K. VetenskAkad. Förh. Stockh.*   **4:** 5–17.

AGARDH, J. G. 1851.   *Species Genera et Ordines Algarum*   **2**(1).   Lundae.

AGARDH, J. G. 1852.   *Species Genera et Ordines Algarum*   **2**(2).   Lundae.

AGARDH, J. G. 1876.   *Species Genera et Ordines Algarum*   **3**(1).   Lundae.

AGARDH, J. G. 1892.   Analecta algologica.   *Acta Univ. lund.* Afd. 2 **28**(6): 1–182.

AGARDH, J. G. 1896.   Analecta algologica Cont. III.   *Acta Univ. lund.* Afd. 2 **32**(2): 1–140.

AGARDH, J. G. 1899.   Analecta algologica Cont. V.   *Acta Univ. lund.* Afd. 2 **35**(4): 1–160.

ALEEM, A. A. 1955.   Marine fungi of the west-coast of Sweden.   *Ark. Bot.* N.S. **3:** 1–33.

ANON. 1860.   Algae marinae siccatae.   *Bot. Zeit.*   **18:**20.

ANON. 1952.   *Flora of Devon* II.   1, The marine algae. Torquay.

ARDISSONE, F. 1893.   Nota alla phycologia Mediterranea.   *Rc. Ist. lomb. Sci. Lett. ser.* 2, **26:** 674–690

ARDRE, F. 1970.   Contrîbution à l'étude des algues marines du Portugal.   *Port. Acta biol.* sér. B **10:** 137–555 (reprint 1–423).

ARDRE, F. 1977. Sur le cycle du *Schizymenia dubyi* (Chauvin ex Duby) J. Agardh (Némastomacée, Gigartinale).   *Rev. algol.* N.S. **12:** 73–86.

ARDRE, F. 1980.   Observations sur le cycle de développement du *Schizymenia dubyi* (Rhodophycée, Gigartinale) en culture, et remarques sur certains genera de Némastomacées.   *Cryptogamie Algol.* **1:** 111–140.

ARDRE, F. & GAYRAL, P. 1961.   Quelques *Grateloupia* de l'Atlantique et du Pacifique.   *Rev. algol.* N.S. **6:** 38–48.

ARESCHOUG, J. E. 1875.   Observationes phycologicae III.   *Nova Acta R. Soc. Scient. upsal.* ser. 3 **10:** 1–36.

AUSTIN, A. H. 1960.   Observations on the growth, fruiting and longevity of *Furcellaria fastigiata* (L.) Lamour.   *Hydrobiologia* **15:** 193–207.

BAARDSETH, E. & TAASEN, J. P. 1973.   *Navicula dumontiae* sp. nov., an endophytic diatom inhabiting the mucilage of *Dumontia incrassata* (Rhodophyceae).   *Norw. J. Bot.* **20:** 80–87.

BARTON, E. S. 1891.   On the occurrence of galls in *Rhodymenia palmata* Grev.   *J. Bot., Lond.* **29:** 65–68.

BATTERS, E. A. L. 1890.   A list of the marine algae of Berwick-on-Tweed.   *Hist. Berwicksh. Nat. Club* **12:** 221–392 (reprint 1–172).

BATTERS, E. A. L. 1895.   On some new British marine algae.   *Ann. Bot.* **9:** 307–320.

BATTERS, E. A. L. 1896.   New or critical British marine algae.   *J. Bot., Lond.* **34:** 384–390.

BATTERS, E. A. L. 1900.   New or critical British marine algae.   *J. Bot., Lond.* **38:** 369–379.

BATTERS, E. A. L. 1902.   A catalogue of the British marine algae.   *J. Bot., Lond.* **40** (Suppl.): 1–107.

BATTERS, E. A. L. 1906.   New or critical British marine algae.   *J. Bot., Lond.* **44:** 1–3.

BEAUVOIS, A. M. F. J. PALISOT DE 1812.   *Essai d'une nouvelle Agrostographie.*   Paris.

BELSHER, T. & MARCOT, J. 1975.   Recherches sur le genre *Peyssonnelia* (Rhodophycées), VI. Étude d'une population de *Peyssonnelia atropurpurea* de Roscoff.   *Cah. Biol. mar.* **16:** 395–413.

BERKELEY, M. J. 1833.   *Gleanings of British algae.*   London.

BERT, J. J. 1965.   Sur la structure et le développement de l'appareil reproducteur femelle de *Dilsea carnosa* (Schmidel) Kuntze et la position systématique du genre *Dilsea.*   *C.r. hebd. Séanc. Acad. Sci., Paris* **261:** 2702–2704.

BERT, J. J. 1967.   Étude des *Callophyllis* (Rhodophycées, Cryptonémiales) des côtes de France.   *Rev. gén. Bot.* **67:** 5–29.

BLIDING, C. 1928.   Studien über die Florideenordnung Rhodymeniales.   *Acta Univ. lund. Ny Följd* Avd. 2 **24**(3): 1–74.

BLUNDEN, G., FARNHAM, W. F., JEPHSON, N., BARWELL, C. J., FENN, R. H. & PLUNKETT, B.A. 1981. The composition of maerl beds of economic interest in northern Brittany, Cornwall and Ireland.   *Proc intl Seaweed Symp.* **10:** 651–656.

BOILLOT, A. 1961.   Recherches sur le mode de développement des spores et la formation de la fronde adulte chez les Champiacées (Rhodophycées, Rhodymeniales).   *Rev. gén. Bot.* **68:** 686–719.

BONEY, A. D. 1975.   Mucilage sheaths of spores of red algae.   *J. mar. biol. Ass. U.K.* **55:** 511–518.

BØRGESEN, F. 1929.   Marine algae from the Canary Islands especially from Teneriffe and Gran Canaria. III, Rhodophyceae. 2, Cryptonemiales, Gigartinales and Rhodymeniales.   *Biol. Meddr* **8:** 1–97.

BØRGESEN, F. 1935.   A list of marine algae from Bombay.   *Biol. Meddr* **12**(2): 1–64.

BORNET, E. 1892.   Les algues de P.-K.-A. Schousboe.   *Mém. Soc. natn. Sci. nat. math. Cherbourg* **28:** 165–376.

BORY DE ST VINCENT, J. B. G. M. 1828.   Cryptogamie, pp. 97–200   *in* Duperrey, L. I., *Voyage autour du monde . . . La Coquille . . . 1822–25.*   Paris.

BOUDOURESQUE, C.-F. & ARDRE, F. 1971.   Recherches sur le genre *Peyssonnelia* (Rhodophycées), II. Présence de *Peyssonnelia rosa-marina* Boud. et Den. au Portugal.   *Pubbl. Staz. zool. Napoli* **39:** 107.

BOUDOURESQUE, C. F. & DENIZOT, M. 1975.   Révision du genre *Peyssonnelia* (Rhodophyta) en Méditerranée.   *Bull. Mus. Hist. nat. Marseille* **35:** 7–92.

BREBNER, G. 1895.   On the origin of the filamentous thallus of *Dumontia filiformis.*   *J. Linn. Soc., Bot.* **30:** 436–443.

BREBNER, G. 1896.   Algological notes.   *J. mar. biol. Ass. U.K.* **4:** 286–288.

BRESSAN, G. 1972.   Osservazioni sugli stadi iniziali della morfogenesi in cultura di alcune species del genere *Peyssonnelia* Decaisne.   *Giorn. Bot. ital.* **106:** 231–244.

BUFFHAM, T. H. 1888.   On the reproductive organs, especially the antheridia, of some of the Florideae.   *J. Quekett microsc. Club* ser. 2, **3:** 257–266.

CABIOCH, J. 1975.   Le *Rhodophysema feldmannii* nov. sp. et les *Rhodophysema* (Rhodophycées, Cryptonemiales ?) de la région de Roscoff.   *Botaniste* **57:** 105–118.

CABIOCH, J. & GUIRY, M. D. 1976.   *Halosacciocolax kjellmanii* Lund, Rhodophycée parasite nouvelle sur les côtes de France.   *Trav. Stn biol. Roscoff* N.S. **23:** 27–29.

CHANG, C. F. & XIA, B. M. 1978.   A new species of *Gastroclonium* from the Xisha Islands, Guangdong Province, China.   *Oceanologia Limnologia sin.* **9:** 209–214.

CHAPMAN, V. J. & PARKINSON, P. G. 1974.   *The marine algae of New Zealand.*   III, Rhodophyceae. Issue 3, Cryptonemiales.   Lehre.

CHAUVIN, F. J. 1831.   *Algues de la Normandie,* 6.   Caen.

CHEMIN, E. 1927.   Action des bactéries sur quelques algues rouges.   *Bull. Soc. bot. Fr.* **74:** 441–451.

CHEMIN, E. 1937.   Rôle des bactéries dans la formation des galles.   *Annls Sci. nat.* sér. 10 Bot. **19:** 61–71.

CODOMIER, L. 1969.   Systématique, morphologie et anatomie de l'espèce *Kallymenia microphylla* J. Agardh.   *Proc. intl Seaw. Symp.* **6:** 107–121.

CODOMIER, L. 1971.   Recherches sur les *Kallymenia* (Cryptonémiales, Kallymeniacées). I. Les espèces méditeranéenes de *Kallymenia.*   *Vie Milieu* sér. A **22:** 1–54.

CODOMIER, L. 1972.   Sur le développment comparé des spores des Sebdéniacées (Gigartinales) et des

Cryptonémiacées (Cryptonémiales). *C.r. hebd. Séanc. Acad. Sci., Paris. Sér.* D. **275**: 747–749.

CODOMIER, L. 1973. Sur le développement des spores et la formation du thalle rampant de *Kallymenia microphylla* J. Ag. (Rhodophyceae, Cryptonémiales). *G. Bot. ital.* **107**: 269–280.

CODOMIER, L. 1974. Recherches sur la structure et le développement des *Halymenia* C. Ag. (Rhodophycées, Cryptonémiales) des côtes de France et de la Méditerranée. *Vie Milieu* sér. A **24**: 1–42.

CODOMIER, L. 1974a. Recherches sur les *Kallymenia* (Cryptonémiales, Kallymeniacées). II. Développement des spores et morphogenèse. *Vie Milieu* sér. A **24**: 369–388.

COTTON, A. D. 1912. Marine Algae, *in* Praeger, R. L., A biological survey of Clare Island in the county of Mayo, Ireland and of the adjoining district. *Proc. R. Ir. Acad.* **31** sect. 1(15): 1–178.

CROUAN, P. L. & H. M. 1835. Observations microscopiques sur le genre *Mesogloia* Agardh. *Annls Sci. nat.* sér. 2 Bot. **3**: 98–100.

CROUAN, P. L. & H. M. 1844. Observations sur le genre *Peyssonnelia* Dne. *Annls Sci. nat.* sér. 3 Bot. **2**: 367–368.

CROUAN, P. L. & H. M. 1858. Note sur quelques algues marines nouvelles de la rade de Brest. *Annls Sci. nat.* sér. 4 Bot. **9**: 69–75.

CROUAN, P. L. & H. M. 1859. Notice sur quelques espèces et genres nouveaux d'algues marines de la rade de Brest. *Annls Sci. nat.* sér 4 Bot. **12**: 288–292.

CROUAN, P. L. & H. M. 1867. *Florule du Finistère.* Paris & Brest.

CULLINANE, J. P. & WHELAN, P. M. 1981. Ecology, distribution and seasonality of *Cryptonemia hibernica* Guiry et Irvine on the south coast of Ireland. *Proc. intl Seaweed Symp.* **10**: 259–264.

DANGEARD, P. J. L. 1949. Les algues marines de la côte occidentale du Maroc. *Botaniste* **34**: 89–189.

DAWSON, E. Y. 1941. A review of the genus *Rhodymenia*, with descriptions of new species. *Allan Hancock Pacif. Exped.* **3**: 123–180.

DAWSON, E. Y. 1952. Marine red algae of Pacific Mexico. I, Bangiales to Corallinaceae subf. Corallinoideae. *Allan Hancock Pacif. Exped.* **17**: 1–238.

DAWSON, E. Y., ACLETO, C. & FOLDVIK, N. 1964. The seaweeds of Peru. *Beih. nov. Hedwigia* **13**: 1–111.

DECAISNE, J. 1841. Plantes de l'Arabie heureuse receuillies par M. P.-E. Botta. *Archs Mus. natn. hist. nat. Paris* **2**: 89–199.

DECAISNE, J. 1842. Essai sur une classification des algues et des polypiers calcifères de Lamouroux. *Annls Sci. nat.* sér. 2 Bot. **17**: 297–380.

DECEW, T. C. 1981. A sexual life history in *Rhodophysema* (Rhodophyceae), a reinterpretation. *J. Phycol.* **17** (Suppl.): 4.

DECEW, T. C. & WEST, J. W. 1981. Investigations on the life histories of three *Farlowia* species (Rhodophyta: Cryptonemiales, Dumontiaceae) from Pacific North America. *Phycologia* **20**: 342–351.

DECEW, T. C. & WEST, J. W. 1982. A sexual life history in *Rhodophysema* (Rhodophyceae): a re-interpretation. *Phycologia* **21**: 67–74.

DECEW, T. C., WEST, J. A. & GANESAN, E. K. 1981. The life histories and developmental morphology of two species of *Gloiosiphonia* (Rhodophyta: Cryptonemiales, Gloiosiphoniaceae) from the Pacific coast of North America. *Phycologia* **20**: 415–423.

DENIZOT, M. 1968. *Les algues floridées encroûtantes (à l'exclusion des Corallinacées).* Paris.

DESVAUX, N. A. 1809. *Champia* Desv., *Ulva* L., *Mertensia* Thunb. *J. Bot. Paris* **1**: 240–246.

DIXON, P. S. 1959. Taxonomic and nomenclatural notes on the Florideae. *Bot. Notiser* **112**: 339–352.

DIXON, P. S. 1961. List of marine algae collected in the Channel Islands. *Brit. phycol. Bull.* **2**: 71–80.

DIXON, P. S. 1961a. The occurrence of tetrasporangia and carposporophytes on the same thallus in *Euthora cristata* (L. ex Turn.) J. Ag. *Can. J. Bot.* **39**: 541–543.

DIXON, P. S. 1964. Taxonomic and nomenclatural notes on the Florideae, IV. *Bot. Notiser* **117**: 56–78.

DIXON, P. S. 1964a. The British marine Rhodophyta: some further deletions and corrections. *Brit. phycol. Bull.* **2**: 393–394.

DIXON, P. S. 1966. On the form of the thallus in the Florideophyceae, pp. 45–63 *in* Cutter, E. (Ed.) *Trends in plant morphogenesis.* London.

DIXON, P. S. & IRVINE, L. M. 1970. Miscellaneous notes on algal taxonomy and nomenclature, III. *Bot. Notiser* **123**: 474–487.

DIZERBO, A. H. 1969.   Les limites géographiques de quelques algues marines du massif armoricain. *Proc. intl Seaweed Symp.* **6**: 141–149.

DUBY, J. E. 1830.   *Botanicon gallicum* ed. 2; **2** Paris.

DUFOUR, L. 1864.   Elenco delle Alghe della Líguria.   *Comment. Soc. crittogam. Ital.* **2**: 28–75.

EDELSTEIN, T. 1970.   The life history of *Gloiosiphonia capillaris* (Hudson) Carmichael.   *Phycologia* **9**: 55–59.

EDELSTEIN, T. 1972.   *Halosacciocolax lundii,* sp. nov., a new red alga parasitic on *Rhodymenia palmata* (L.) Grev.   *Br. phycol. J.* **7**: 249–253.

EDELSTEIN, T. & MCLACHLAN, J. 1971.   Further observations on *Gloiosiphonia capillaris* (Hudson) Carmichael in culture.   *Phycologia* **10**: 215–219.

EDELSTEIN, T. & MCLACHLAN, J. 1977.   On *Choreocolax odonthaliae* Levring (Cryptonemiales, Rhodophyceae).   *Phycologia* **16**: 287–293.

EKMAN, F. L. 1857.   *Bidrag till Kännedomen af Skandinaviens Hafsalger.*   Stockholm.

ENGLER, A. 1892.   *Syllabus der Vorlesungenüber specielle und medicinisch-pharmaceutische Botanik.* Grosse Ausgabe. Berlin.

ERCEGOVIC, A. 1956.   Famille des Champiacées (Champiaceae) dans l'Adriatique moyenne.   *Acta adriat.* **8**(2): 1–64.

EVANS, L. V., CALLOW, J. A. & CALLOW, M. E. 1973.   Structural and physiological studies on the parasitic red alga *Holmsella.*   *New Phytol.* **72**: 393–402.

EVANS, L. V., CALLOW, J. A. & CALLOW, M. E. 1978.   Parasitic red algae: an appraisal, pp. 87–109 *in* Irvine, D. E. G. & Price, J. H. (Eds) *Modern approaches to the taxonomy of red and brown algae.* Systematics Association Special Volume **10**. London.

EVANS, L. V., CALLOW, M. E. & CALLOW, J. A. 1981.   Host/parasite relationships in seaweeds.   *Proc. intl Seaweed Symp.* **8**: 167–171.

FARLOW, W. G. 1881.   *Marine algae of New England and adjacent coasts.*   Washington.

FARNHAM, W. F. 1980.   Studies on aliens in the marine flora of southern England,   *in* Price, J. H., Irvine, D. E. G. & Farnham, W. F. (Eds) *The shore environment* **2**: 875–914.   Systematics Association Special Volume **17b**. London.

FARNHAM, W. F. & IRVINE, L. M. 1968.   Occurrence of unusually large plants of *Grateloupia* in the vicinity of Portsmouth.   *Nature, Lond.* **219**: 744–746.

FARNHAM, W. F. & IRVINE, L. M. 1973.   The addition of a foliose species of *Grateloupia* to the British marine flora.   *Br. phycol. J.* **8**: 208–209.

FARNHAM, W. F. & JONES, E. B. G. 1973.   Présence à Roscoff du *Chadefaudia gymnogongri* (Ascomycète), parasite du *Grateloupia filicina* (Rhodophycée, Cryptonémiale).   *Trav. Stn biol. Roscoff* N.S. **20**(31): 8.

FEE, A. L. A. 1824.   *Essai sur les cryptogames des écorces exotiques officinales.*   Paris.

FELDMANN, G. 1967.   Le genre *Cordylecladia* J. Ag. (Rhodophycées, Rhodyméniales) et sa position systématique.   *Rev. gén. Bot.* **74**: 357–375.

FELDMANN, J. 1939.   Les algues marines de la côte des Albères. IV, Rhodophycées.   *Rev. algol.* **11**: 247–330.

FLETCHER, R. L. 1975.   The life history of *Rhodophysema georgii* in laboratory culture.   *Mar. Biol. Berlin* **31**: 299–304.

FLETCHER, R. L. 1977.   The life history of *Rhodophysema elegans* in laboratory culture.   *Mar. Biol. Berlin* **40**: 291–298.

FORAN, C. F. & Guiry, M. D. 1983.   The life history in culture of isolates of *Lomentaria orcadensis* (Harv.) F. S. Collins ex W. R. Taylor (Rhodophyta) from Ireland and Scotland.   *Br. phycol. J.* **18**: in press.

FRITSCH, F. E. 1945.   *Structure and reproduction of the algae* **2**. Cambridge.

GAILLON, B. 1828.   Thalassiophytes,   *in Dictionnaire des Sciences naturelles* **53**: 350–406.   Paris & Strasbourg.

GANESAN, E. K. & WEST, J. A. 1975.   Culture studies on the marine red alga *Rhodophysema elegans* (Cryptonemiales, Peysonneliaceae).   *Phycologia* **14**: 161–166.

GARBARY, D. 1979.   Numerical taxonomy and generic circumscription in the Acrochaetiaceae (Rhodophyta).   *Botanica mar.* **22**: 477–492.

GAYRAL, P. 1958. *Algues de la Côte Atlantique Marocaine.* Rabat.

GEPP, A. & E. S. 1906. Some marine algae from New South Wales. *J. Bot., Lond.* **44**: 249–261.

GMELIN, S. G. 1768. *Historia fucorum.* Petropoli.

GOFF, L. J. 1976. The biology of *Harveyella mirabilis.* V, Host responses to parasite infection. *J. Phycol.* **12**: 313–328.

GOFF, L. J. & COLE, K. 1973. The biology of *Harveyella mirabilis* (Cryptonemiales, Rhodophyceae). I, Cytological investigations of *Harveyella mirabilis* and its host *Odonthalia floccosa.* *Phycologia* **12**: 237–245.

GOFF, L. J. &COLE, K. 1975. The biology of *Harveyella mirabilis* (Cryptonemiales, Rhodophyceae). II, Carposporophyte development as related to the taxonomic affiliation of the parasitic red alga, *Harveyella mirabilis.* *Phycologia* **14**: 227–238.

GOFF, L. J. & COLE, K. 1976. The biology of *Harveyella mirabilis* (Cryptonemiales, Rhodophyceae). III, Spore germination and subsequent development within the host *Odonthalia floccosa* (Ceramiales, Rhodophyceae). *Can. J. Bot.* **54**: 268–280.

GOFF, L. J. & COLE, K. 1976a. The biology of *Harveyella mirabilis* (Cryptonemiales, Rhodophyceae). IV, Life history and phenology. *Can. J. Bot.* **54**: 281–291.

GOODENOUGH, S. & WOODWARD, T. J. 1797. Observations on the British *Fuci*, with particular descriptions of each species. *Trans. Linn. Soc. Lond.* **3**: 84–235.

GOOR, A. C. S. VAN 1923. Die Holländischen Meeresalgen. *Verh. K. Akad. Wet.* **23**(2):1–232.

GRAY, S. F. 1821. *A natural arrangement of British plants* **1**. London.

GREVILLE, R. K. 1826. Some account of a collection of cryptogamic plants from the Ionian Islands. *Trans. Linn. Soc. Lond.* **15**: 335–348.

GREVILLE, R. K. 1828. *Scottish cryptogamic flora* **6**: 301–360. Edinburgh.

GREVILLE, R. K. 1830. *Algae Britannicae.* Edinburgh.

GUIRY, M. D. 1974. The occurrence of the red algal parasite *Halosacciocolax lundii* Edelstein in Britain. *Br. phycol. J.* **9**: 31–35.

GUIRY, M. D. 1974a. A preliminary consideration of the taxonomic position of *Palmaria palmata* (Linnaeus) Stackhouse = *Rhodymenia palmata* (Linnaeus) Greville. *J. mar. biol. Ass. U.K.* **54**: 509–528.

GUIRY, M. D. 1975. *Halosacciocolax kjellmanii* Lund parasitic on *Palmaria palmata* forma *mollis* (S. et G.) Guiry in the eastern North Pacific. *Syesis* **8**: 113–117.

GUIRY, M. D. 1977. Studies on marine algae of the British Isles. 10, The genus *Rhodymenia.* *Br. phycol. J.* **12**: 385–425.

GUIRY, M. D. 1978. The importance of sporangia in the classification of the Florideophyceae, pp. 111–144 *in* Irvine, D. E. G. & Price, J. H. (Eds) *Modern approaches to the taxonomy of red and brown algae.* Systematics Association Special Volume **10**. London.

GUIRY, M. D., CULLINANE, J. P. & WHELAN, P. 1979. Notes on Irish marine algae. 3, New records of Rhodophyta from the Wexford coast. *Irish Nat. J.* **19**: 304–307.

GUIRY, M. D. & HOLLENBERG, G. J. 1975. *Schottera* gen. nov. and *Schottera nicaeensis* (Lamour. ex Duby) comb. nov. (= *Petroglossum nicaeense* (Lamour. ex Duby) Schotter) in the British Isles. *Br. phycol. J.* **10**: 149–164.

GUIRY, M. D. & IRVINE, D. E. G. 1981. A critical reassessment of infraordinal classification in the Rhodymeniales. *Proc. intl Seaweed Symp.* **8**: 106–111.

GUIRY, M. D. & IRVINE, L. M. 1974. A species of *Cryptonemia* new to Europe. *Br. phycol. J.* **9**: 225–237.

GUIRY, M. D. & MAGGS, C. A. 1982. The life-history of *Meredithia microphylla* (J. Ag.) J. Ag. (Rhodophyta) in culture. *Br. phycol. J.* **17**: 232–233.

GUIRY, M. D. & MAGGS, C. A. 1982a. The morphology and life history of *Dermocorynus montagnei* Crouan frat. (Halymeniaceae; Rhodophyta) from Ireland. *Br. phycol. J.* **17**: 215–228.

HARDING, J. P. 1954. The copepod *Thalestris rhodymeniae* (Brady) and its nauplius parasitic in the seaweed *Rhodymenia palmata* (L.) Grev. *Proc. zool. Soc. Lond.* **124**: 153–161.

HARVEY, W. H. 1844. Description of a minute alga from the coast of Ireland. *Ann. Mag. nat. Hist.* **14**: 27–28.

HARVEY, W. H. 1846. *Phycologia Britannica* pl. i–lxxii. London.

HARVEY, W. H. 1849. *A manual of the British marine algae* London.

HARVEY, W. H. 1850. *Phycologia Britannica* pl. cclix–cccliv. London.

HARVEY, W. H. 1851. *Phycologia Britannica* pl. ccclv–ccclx. London.

HARVEY, W. H. 1853. *Nereis Boreali–Americana* 2 Rhodospermae. Washington.

HAUCK, F. 1885. Die Meeresalgen Deutschlands und Oesterreichs, *in* Rabenhorst, L., *Kryptogamen-Flora von Deutschland, Oesterreich und der Schweiz* ed. 2. 2 Leipzig.

HEYDRICH, F. 1903. Über *Rhododermis* Crouan. *Beih. bot. Zbl.* **14:** 243–246.

HEYDRICH, F. 1905. *Polystrata*, eine Squamariacee aus den Tropen. *Ber. dt. Bot. Ges.* **23:** 30–36.

HOEK, C. VAN DEN 1982. The distribution of benthic marine algae in relation to the temperature regulation of their life histories. *Biol. J. Linn. soc.* **18:** 81–114.

HOLLENBERG, G. J. 1940. New marine algae from Southern California, I. *Am. J. Bot.* **27:** 868–877.

HOLLENBERG, G. J. 1971. Phycological notes. VI, New records, new combinations and noteworthy observations concerning marine algae of California. *Phycologia* **10:** 281–289.

HOLLENBERG, G. J. & ABBOTT, I. A. 1965. New species and new combinations of marine algae from the region of Monterey, California. *Can. J. Bot.* **43:** 1177–1188.

HOLMES, E. M. 1883. *Rhodymenia palmetta*, Var. *nicaeensis*. *J. Bot., Lond.* **21:** 289–290.

HOLMES, E. M. 1907. *Callymenia Larteriae*, n. sp. *J. Bot., Lond.* **45:** 85–86.

HOMMERSAND, M. & OTT, D. W. 1970. Development of the carposporophyte of *Kallymenia reniformis* (Turner) J. Agardh. *J. Phycol.* **6:** 322–331.

HOOKER, W. J. 1833. *British flora* **5.** London.

HOOPER, R. G. & SOUTH, G. R. 1974. A taxonomic appraisal of *Callophyllis* and *Euthora* (Rhodophyta). *Br. phycol. J.* **9:** 423–428.

HOWE, M. A. 1914. The marine algae of Peru. *Mem. Torrey bot. Club* **15:** 1–185.

HUDSON, W. 1762. *Flora Anglica.* London.

HUDSON, W. 1778. *Flora Anglica* ed. 2. London.

IRVINE, D. E. G., GUIRY, M. D., TITTLEY, I. & RUSSELL, G. 1975. New and interesting marine algae from the Shetland Isles. *Br. phycol. J.* **10:** 57–71.

IRVINE, L. M. & DIXON, P. S. 1982. The typification of Hudson's algae: a taxonomic and nomenclatural reappraisal. *Bull. Br. Mus. nat. Hist.* (Bot.) **10:** 91–105.

JAO, C. C. 1936. New Rhodophyceae from Woods Hole. *Bull. Torrey bot. Club* **63:** 237–258.

JONES, W. E. 1962. The identity of *Gracilaria erecta* (Grev.) Grev. *Br. phycol. Bull.* **2:** 140–144.

KAIN, J. M. 1960. Direct observations on some Manx sublittoral algae. *J. mar. biol. Ass. U.K.* **39:** 609–630.

KILAR, J. A. & MATHIESON, A. C. 1978. Ecological studies of the annual red alga *Dumontia incrassata* (O. F. Müller) Lamouroux. *Botanica mar.* **21:** 423–437.

KNIGHT, M. & PARKE, M. W. 1931. Manx algae. An algal survey of the south end of the Isle of Man. *L.M.B.C. Mem. typ. Br. mar. Pl. Anim.* **30:** 1–155.

KRAFT, G. T. 1977. The morphology of *Grateloupia intestinalis* from New Zealand, with some thoughts on generic criteria within the family Cryptonemiaceae (Rhodophyta). *Phycologia* **16:** 43–51.

KRISTIANSEN, A. 1972. A seasonal study of the marine algal vegetation in Tuborg harbour, the Sound, Denmark. *Bot. Tidsskr.* **67:** 201–244.

KUCKUCK, P. 1897. Bemerkungen zur marinen Algenvegetation von Helgoland II. *Wiss. Meeresunters. (Helgol.)* N.F. **2:** 371–400.

KUGRENS, P. & WEST, J. A. 1975. The ultrastructure of an alloparasitic red alga *Choreocolax polysiphoniae*. *Phycologia* **12:** 175–187.

KUNTZE, C. E. O. 1891. *Revisio Generum Plantarum* **2:** 375–1011. Leipzig.

KUNTZE, C. E. O. 1893. *Revisio Generum Plantarum* **3:** 1–576. Leipzig.

KÜTZING, F. T. 1843. *Phycologia Generalis.* Leipzig.

KÜTZING, F. T. 1849. *Species Algarum.* Leipzig.

KÜTZING, F. T. 1866. *Tabulae Phycologicae* **16.** Nordhausen.

KYLIN, H. 1907. *Studien über die Algenflora des schwedischen Westküste.* Upsala.

KYLIN, H. 1923. Studien über die Entwicklungsgeschichte der Florideen. *K. svenska VetenskAkad. Handl.* **63**(11): 1–139.

REFERENCES   109

KYLIN, H. 1928.   Entwicklungsgeschtliche Florideenstudien.   *Acta Univ. lund.* Ny Földj, Avd 2. **24**(4): 1–127.

KYLIN, H. 1930.   Über die Entwicklungsgeschichte der Florideen.   *Acta Univ. lund.* Ny Földj, Avd 2. **26**(6): 1–14.

KYLIN, H. 1931.   Die Florideenordnung Rhodymeniales.   *Acta Univ. lund.* Ny Földj, Avd 2. **27**(11): 3–48.

KYLIN, H. 1956.   *Die Gattungen der Rhodophyceen.*   Lund.

LAMARCK, J. B. P. A. DE M. DE & POIRET, J. L. M. 1808.   *Encyclopédie Méthodique.*   *Botanique* **8.** Paris.

LAMI, R. 1940.   Sur les epiphytes hiverneaux des stipes de Laminaires et sur deux *Rhodochorton* qui s'observent dans le région de Dinard.   *Bull. Lab. marit. Dinard* **22:** 47–60.

LAMOUROUX, M. J. V. F. 1805.   *Dissertations sur Plusieurs Espèces de Fucus peu connue ou nouvelles; avec leur Description en Latin et en Français.*   Agen & Paris.

LAMOUROUX, M. J. V. F. 1813.   Essai sur les genres de la famille des Thalassiophytes non articulées. *Annls Mus. Hist. nat. Paris* **20:** 21–47; 115–139; 267–293 [reprint 1–84].

LANDSBOROUGH. D. 1844   On the fructification of *Gloiosiphonia capillaris.*   *Ann. Mag. nat. Hist.* **14:** 240–241.

LAWSON, G. R., JOHN, D. M. & PRICE, J. H. 1975.   The marine algal flora of Angola: its distribution and affinities.   *Bot. J. Linn. Soc.* **70:** 307–324.

LEE, I. K. 1978.   Studies on Rhodymeniales from Hokkaido.   *J. Fac. Sci. Hokkaido Univ.* ser. V (Botany) **11:** 1–194.

LEE, I. K. & KUROGI, M. 1978   *Neohalosacciocolax aleutica* gen. et sp. nov. (Rhodophyta), parasitic on *Halosaccion minjaii* I. K. Lee from the North Pacific.   *Br. phycol. J.* **13:** 131–139.

LE JOLIS, A. 1863.   Liste des algues marines de Cherbourg.   *Mém. Soc. natn. Sci. nat. math. Cherbourg* **10:** 6–168.

LEVRING, T. 1935.   Untersuchungen aus dem Oresund. XIX, Zur Kenntnis der Algenflora von Kullen an der schwedischen Westküste.   *Acta Univ. lund.* Ny Földj, Avd 2. **31**(4): 1–64.

LEVRING, T., HOPPE, H. A. & SCHMID, O. J. 1969.   *Marine algae. A survey of research and utilization.* Hamburg.

L'HARDY-HALOS, M.-TH. 1970.   *Rhodymenia phylloïdes* nov. sp. (Rhodophycées, Rhodyméniale) nouvelle espèce des côtes Bretagne.   *Bull. Soc. phycol. Fr.* **15:** 23–30.

L'HARDY-HALOS, M.-TH. 1976.   A propos du *Rhodymenia coespitosella* sp. nov. (Rhodophycée, Rhodyméniale); comparaisons morphologiques, anatomiques et cytologiques.   *Phycologia* **15:** 289–297.

LIGHTFOOT, J. 1777.   *Flora Scotica* **2.**   London.

LINNAEUS, C. 1753.   *Species plantarum* **2.**   Holmiae.

LODGE, S. M. 1948.   Additions to algal records for the Manx region.   *Rep. mar. Biol. Stn Port Erin* **58–60:** 59–62.

LUCAS, J. A. W. 1950.   The algae transported on drifting objects and washed ashore on the Netherlands' coast.   *Blumea* **6:** 527–543.

LUDWIG, C. G. 1757.   *Institutiones historico-physicae Regni Vegetabilis.*   ed. 2.   Lipsiae.

LUND, S. 1959.   The marine algae of East Greenland. I, Taxonomical part.   *Meddr Grønland.* **156**(1): 1–248.

LYNGBYE, H. C. 1819.   *Tentamen Hydrophytologiae Danicae.*   Hafniae.

MAGGS, C. A., FREAMHAINN, M.T. & GUIRY, M. D. 1983.   A study of the marine algae of subtidal cliffs in Lough Hyne (Ine.), Co. Cork   *Proc. R. Ir. Acad*   In press.

MAGGS, C. A. & GUIRY, M. D. 1982.   Notes on Irish marine algae. 5, Preliminary observations on deep water vegetation off West Donegal.   *Ir. Nat. J.* **20:** 357–361.

MAGGS, C. A. & GUIRY, M. D. 1982a.   The life history of *Haematocelis fissurata* Crouan frat. (Rhodophyta: Sphaerococcaceae).   *Br. phycol. J.* **17:** 235.

MAGGS, C. A. & GUIRY, M. D. 1982b.   Morphology, phenology and photoperiodism in *Halymenia latifolia* Kütz. (Rhodophyta) from Ireland.   *Botanica mar.* **25:** 589–599.

MAGGS, C. A., GUIRY, M. D. & IRVINE, L. M. 1983.   The life history in culture of an isolate of *Rhododiscus pulcherrimus* Crouan frat. (Rhodophyta) from Ireland.   *Br. phycol. J.* **18:**   In press.

MAGGS, C. A. & IRVINE, L. M. 1983 *Peyssonnelia immersa* sp. nov. (Cryptonemiales, Rhodophyta) from the British Isles and France, with a survey of infrageneric classification. *Br. phycol. J.* **18:** In press.

MAGNE, F. 1964. Recherches caryologiques chez les Floridées (Rhodophycées). *Cah. Biol. mar.* **5:** 461–671.

MAIRE, R. & CHEMIN, E. 1922. Un nouveau Pyrenomycète marin. *C.r. hebd. Séanc. Acad. Sci., Paris* **175:** 319–321.

MARCOT-COQUEUGNIOT, J. 1980. Recherches sur le genre *Peyssonnelia* (Rhodophyta). XIII, Sur un *Peyssonnelia* du 'complexe *harveyana*'. *Botanica mar.* **23:** 35–39.

MARCOT, J. & BOUDOURESQUE, C.-F. 1976. Recherches sur le genre *Peyssonnelia* (Rhodophyta). VIII, Etude du type de *P. harveyana* J. Agardh. *Bull. Mus. Hist. nat. Marseille* **36:** 5–9.

MARCOT, J. & BOUDOURESQUE, C.-F. 1977. Recherches sur le genre *Peyssonnelia* (Rhodophyta). XI, Sur un *Peyssonnelia* de Corse. *Bull. Mus. Hist. nat. Marseille.* **37:** 109–116.

MARCOT, J., BOUDOURESQUE, C.-F. & VERLAQUE, M. 1977. Recherches sur le genre *Peyssonnelia* (Rhodophycées). IX, Les nemathécies à sporocystes des *Peyssonnelia* de Mediterranée. *Bull. Soc. phycol. Fr.* **22:** 70–78.

MASUDA, M. 1976. Taxonomic notes on *Rhodophysemopsis* gen. nov. (Rhodophyta). *J. Jap. Bot.* **51:** 175–186.

MASUDA, M. & OHTA, M. 1975. The life history of *Rhodophysema georgii* Batters (Rhodophyta, Cryptonemiales). *J. Jap. Bot.* **50:** 1–10.

MASUDA, M. & OHTA, M. 1981. Taxonomy and life history of *Rhodophysema odonthaliae* sp. nov. *Jap. J. Phycol.* **29:** 7–21.

MASUDA, M. & OHTA, M. 1981a. A taxonomic study of *Rhodophysema elegans* (Rhodophyta) from Japan. *Acta phytotax. geobot. Kyoto* **32:** 75–89.

MATHIAS, W. T. 1935. The life history and cytology of *Phloeospora brachiata* Born. *Publs Hartley bot. Labs Lpool Univ.* **13:** 1–24.

MIRANDA, F. 1931. Observaciones sobre Florideas. *Boln R. Soc. esp. Hist. nat.* **31:** 187–196.

MIRANDA, F. 1932. Remarques sur quelques algues marines des côtes de la Manche. *Rev. algol.* **6:** 275–296.

MONTAGNE, J. F. C. 1839. Florula Boliviensis stirpes novae et minus cognitae. Plantae cellulares, *in* Orbigny, A.d', *Voyage dans l'Amérique meridionale* **7**(2): 1–39.

MONTAGNE, J. F. C. 1846. Phyceae, *in* Durieu de Maisonneuve, M. C., *Exploration Scientifique de l'Algérie*. Sciences Naturelles Botanique **1:** 1–197.

MORGAN, K. C., WRIGHT, J. L. C. & SIMPSON, F. J. 1980. Review of chemical constituents of the red alga *Palmaria palmata* (Dulse). *Econ. Bot.* **34:** 27–50.

MOROHOSHI, H. & MASUDA, M. 1980. The life history of *Gloiosiphonia capillaris* (Hudson) Carmichael (Rhodophyceae, Cryptonemiales). *Jap. J. Phycol.* **28:** 81–91.

MÜLLER, O. F. 1775. *Flora Danica* **4.** Havniae.

NEWTON, L. 1931. *A handbook of the British seaweeds*. London.

NORRIS, R. E. 1957. Morphological studies on the Kallymeniaceae. *Univ. Calif. Publs Bot.* **28:** 251–334.

NORTON, T. A. 1970. The marine algae of county Wexford, Ireland. *Br. phycol. J.* **5:** 257–266.

OLTMANNS, F. 1904. *Morphologie und Biologie der Algen*. Jena.

PAPENFUSS, G. F. 1944. Notes on algal nomenclature. III, Miscellaneous species of Chlorophyceae, Phaeophyceae and Rhodophyceae. *Farlowia* **1:** 337–346.

PARKE, M. & DIXON, P. S. 1976. Check-list of British marine algae – third revision. *J. mar. biol. Ass. U.K.* **56:** 527–594.

PARKINSON, P. G. 1981. Proposals to amend the Code. *Taxon* **30:** 274–285.

PARKINSON, P. G. 1981a. *Grateloupia ornata* C. Agardh 1822, nom. rectotyp. prop. *Taxon* **30:** 312–314.

PARKINSON, P. G. 1981b. Remarks on some algal generic names recently proposed for nomenclatural conservation: *Halymenia, Grateloupia, Nemastoma* and *Schizymenia*. *Taxon* **30:** 314–318.

PERCIVAL, E. 1979. The polysaccharides of green, red and brown seaweeds; their basic structure, biosynthesis and function. *Br. phycol. J.* **14:** 103–117.

PEYRIERE, M. 1981. Jonctions cellulaires et synapses des Rhodophycées Floridées. Études de deux

Choreocolacées parasites, *Harveyella mirabilis* et *Holmsella pachyderma. Cryptogamie Algol.* 2: 85–104.

PRINTZ, H. 1926. Die Algenvegetation der Trondhjemsfjorden. *Skr. norske Vidensk-Akad. mat.-nat. Kl.* **1926**(5): 1–274.

PUESCHEL, C. M. 1979. Ultrastructure of tetrasporogenesis in *Palmaria palmata* (Rhodophyta). *J. Phycol.* **15**: 409–424.

PUESCHEL, C. M. 1980. A reappraisal of the cytochemical properties of rhodophycean pit plugs. *Phycologia* **19**: 216–217.

PUESCHEL, C. M. 1982 Ultrastructural observations of tetrasporangia and conceptacles in *Hildenbrandia* (Rhodophyta; Hildenbrandiales). *Br. phycol. J.* **17**: 333–341.

PUESCHEL, C. & COLE, K. 1980. Ultrastructural features bearing on the taxonomic affinities of the Palmariaceae (Rhodophyta). *J. Phycol.* **16**(Suppl.): 34.

PUESCHEL, C. & COLE, K. 1981. Taxonomic implications of an ultrastructural survey of red algal pit plugs. *J. Phycol.* **17**(Suppl.): 14.

PUESCHEL, C. M. & COLE, K. M. 1982. Ultrastructure of pit plugs: a new character for the taxonomy of red algae. *Br. phycol. J.* **17**: 238.

PUESCHEL, C. M. & COLE, K. M. 1982a. Rhodophycean pit plugs: an ultrastructural survey with taxonomic implications. *Am. J. Bot.* **69**: 703–720.

QUIRK, H. M. & WETHERBEE, R. 1980. Structural studies on the host–parasite relationship between the red algae *Holmsella* and *Gracilaria. Micron* **11**: 511–512.

REEDMAN, D. J. & WOMERSLEY, H. B. S. 1976. Southern Australian species of *Champia* and *Chylocladia* (Rhodymeniales: Rhodophyta). *Trans. R. Soc. S. Austr.* **100**: 75–104.

REINKE, J. 1889. Algenflora der westlichen Ostsee, Deutschen Antheils. *Ber. comm. wiss. Untersuch. dt. Meere* **17–19**(1): 1–101.

REINSCH, P. 1875. *Contributiones ad Algologiam et Fungologiam* **1**. Lipsiae.

RICHARDS, H. M. 1891. On the structure and development of *Choreocolax polysiphoniae* Reinsch. *Proc. Am. Acad. Arts Sci.* **26**: 46–63.

RIETEMA, H. 1982. Effects of photoperiod and temperature on macrothallus initiation in *Dumontia contorta. Mar. Ecol. Prog. Ser.* **8**: 187–196

RIETEMA, H. & BREEMAN, A. M. 1982 The regulation of the life history of *Dumontia contorta* in comparison to that of several other Dumontiaceae (Rhodophyta). *Botanica mar.* **25**: 569–576.

RIETEMA, H. & KLEIN, A. W. O. 1981. Environmental control of the life cycle of *Dumontia contorta* (Rhodophyta) kept in culture. *Mar. Ecol. Prog. Ser.* **4**: 23–29.

ROSENVINGE, L. K. 1893. Grønlands havalger. *Meddr Grønland* **3**: 765–981.

ROSENVINGE, L. K. 1917. The marine algae of Denmark. Contributions to their natural history, part II. Rhodophyceae, II (Cryptonemiales). *K. danske Vidensk. Selsk. Skr.* 7 Raekke. Nat. Math. Afd. **7**: 153–284.

ROSENVINGE, L. K. 1931. The marine algae of Denmark. Contributions to their natural history, part IV. Rhodophyceae, IV (Gigartinales, Rhodymeniales, Nemastomatales). *K. danske Vidensk. Selsk. Skr.* 7 Raekke Nat. Math. Afd. **7**: 489–628.

ROTH, A. W. 1797. *Catalecta Botanica* **1**. Leipzig.

ROTH, A. W. 1806. *Catalecta Botanica* **3**. Leipzig.

RUPRECHT, F. J. 1850. *Algae Ochotenses* **1**(2). St Petersburg. [Preprint of RUPRECHT, 1851].

RUPRECHT, F. J. 1851. Tange des Ochotskischen Meeres, *in* Middendorff, A. T. von (Ed.) *Sibirische Reise* Botanik 1(2): 193–435. St Petersburg.

SAYRE, G. 1969. Cryptogamae Exsiccatae. *Mem. N.Y. bot. Gdn* **19**: 57–106.

SCHIFFNER, V. 1916. Studien über Algen des adriatischen Meeres. *Wiss. Meeresunters. (Helgol.)* N.F. **11**: 127–198.

SCHMIDEL, C. C. 1794. *Descriptio Itineris per Helvetiam, Galliam et Germaniam partem.* Erlangae.

SCHMITZ, F. 1889. Systematische Übersicht der bisher bekannten Gattungen der Florideen. *Flora, Jena* **72**: 435–456.

SCIUTO, S., PIATTELLI, M., CHILLEMI, R., FURNARI, G. & CORMACI, M. 1979. The implication of *Haematocelis rubens* J. Ag. in the life history of *Schizymenia dubyi* (Chauvin) J. Agardh (Rhodophyta, Gigartinales): a chemical study. *Phycologia* **18**: 296–299.

SEGAWA, S. 1936. On the marine algae of Susaki, Prov. Izu, and its vicinity, II. *Scient. Pap. Inst. algol. Res. Hokkaido* **1**: 175–197.

SETCHELL, W. A. 1923. *Dumontia filiformis* on the New England coast. *Rhodora* **25**: 33–37.

SETCHELL, W. A. 1923a. Parasitic Florideae, II. *Univ. Calif. Publs Bot.* **10**: 393–396.

SILVA, P. C. 1952. A review of nomenclatural conservation in the algae from the point of view of the type method. *Univ. Calif. Publs Bot.* **25**: 241–324.

SILVA, P. C. 1980. *Names of classes and families of living algae.* Regnum Vegetabile volume **103**. Utrecht & The Hague.

SILVA, P. C. & JOHANSEN, H. W. 1983. Reappraisal of the order Corallinales. *Br. phycol. J.* **18**: in press.

SKOTTSBERG, C. 1923. Botanische ergebnisse der schwedischen expedition nach Patagonien und dem Feuerlande 1907–1909. IX, Marine Algae Rhodophyceae. *K. svenska VetenskAkad. Handl.* **63**(8): 1–70.

SONDER, W. 1852. Plantae Müllerianae. Algae *Linnaea* **25**: 657–703.

SOUTH, G. R. & WHITTICK, A. 1976. Aspects of the life history of *Rhodophysema elegans* (Rhodophyta, Peyssonneliaceae). *Br. phycol. J.* **11**: 349–354.

SPARLING, S. R. 1957. The structure and reproduction of some members of the Rhodymeniaceae. *Univ. Calif. Publs Bot.* **29**: 319–396.

STACKHOUSE, J. 1801. *Nereis Britannica* ed. 1. **3**. Bathoniae & Londini.

STACKHOUSE, J. 1809. Tentamen marino-cryptogamicum. *Mem. Soc. Nat. Moscou* **2**: 50–97.

STACKHOUSE, J. 1816. *Nereis Britannica* ed. 2. **2**. Bathoniae & Londini.

STEELE, R. L. & THURSBY, G. B. 1981. Development of a bioassay using the life cycle of *Champia parvula* (Rhodophyta). *Phycologia* **20**: 114.

STRÖMFELT, H. F. G. 1886. Einige für die Wissenschaft neue Meeresalgen aus Island. *Bot. Zbl.* **26**: 172–173.

STURCH, H. H. 1924. On the life-history of *Harveyella pachyderma* and *H. mirabilis. Ann. Bot.* **38**: 27–42.

STURCH, H. H. 1926. *Choreocolax Polysiphoniae* Reinsch. *Ann. Bot.* **40**: 585–605.

SVEDELIUS, N. 1937. The apomeiotic tetrad division in *Lomentaria rosea. Symb. bot. upsal.* **2**(2): 1–54.

TAYLOR, W. R. 1937a. Notes on north Atlantic marine algae, 1. *Pap. Mich. Acad. Sci.* **22**: 225–233.

TAYLOR, W. R. 1937. *Marine algae of the northeastern coast of North America.* Ann Arbor.

TITTLEY, I. & PRICE, J. H. 1977. An atlas of the seaweeds of Kent. *Trans. Kent Fld Club* **7**: 1–80.

TOKIDA, J. 1934. Phycological observations. *Trans. Sapporo nat. Hist. Soc.* **13**: 196–202.

TOKIDA, J. 1954. The marine algae of southern Saghalien. *Mem. Fac. Fish Hokkaido Univ.* **2**: 1–264.

TREVISAN, V. B. A. 1843. Synopsis generum algarum. *Atti Riun. Sci. ital.* **4**: 332–335.

TREVISAN, V. B. A. 1848. *Saggio di una Monografia delle Alghe Coccotalle.* Padova.

TURNER, D. 1801. Descriptions of four new species of *Fucus. Trans. Linn. Soc. Lond.* **6**: 125–136.

TURNER, D. 1808. Fuci **1**. London.

TURNER, D. 1809. Fuci **2**. London.

VAN DER MEER, J. P. 1981. Sexual reproduction in the Palmariaceae. *Proc. intl Seaweed Symp.* **10**: 191–196.

VAN DER MEER, J. P. 1981a. The life history of *Halosaccion ramentaceum. Can. J. Bot.* **59**: 433–436.

VAN DER MEER, J. P. & CHEN, L. C.-M. 1979. Evidence for sexual reproduction in the red algae *Palmaria palmata* and *Halosaccion ramentaceum. Can. J. Bot.* **57**: 2452–2459.

VAN DER MEER, J. P. & TODD, E. R. 1980. The life history of *Palmaria palmata.* A new type for the Rhodophyta. *Can. J. Bot.* **58**: 1250–1256.

VERLAQUE, M. 1978. Recherches sur le genre *Peyssonnelia* (Rhodophycées). X, Présence de *Peyssonnelia codana* (Rosenvinge) Denizot en méditerranée. *Giorn. Bot. ital.* **112**: 29–39.

WEST, J. A. 1970. The life history of *Rhodochorton concrescens* in culture. *Br. phycol. J.* **5**: 179–186.

WHELDEN, R. M. 1928. Observations on the red alga *Dumontia filiformis. Me Nat.* **8**: 121–130.

WILLDENOW, C. L. 1804. Mertensia, ett nytt slägte af Ormbunkarne. *K. svenska VetenskAkad. Handl.* N.S. **25**: 165–170.

WITHERING, W. 1796. *An arrangement of British plants.* ed. 3. **4**. Birmingham & London.

WULFEN, F. X. 1789. Plantae rariores Carinthiacae, *in* Jacquin, N. J., *Collecteana ad Botanicam, Chemicam et Historiam Naturalem.* **3**: 3–166. Vienna.

WYNNE, M. J. & TAYLOR, W. R. 1973. The status of *Agardhiella tenera* and *Agardhiella baileyi* (Rhodophyta, Gigartinales). *Hydrobiologia* **43**: 93–107.

ZANARDINI, G. 1842 [1841] Synopsis algarum in mari adriatico. *Memorie Accad. Sci. Torino* ser. 2 **4**: 105–255.

ZANARDINI, G. 1843. *Saggio di Classificazione Naturale delle Ficee.* Venezia.

ZANARDINI, G. 1844. Rivista critica delle Corallinee (o Polypai calciferi di Lamouroux). *Atti R. Ist. veneto Sci.* **3**: 186–188.

ZANARDINI, G. 1863. Iconographia phycologica adriatica ossia scelta di ficee nuove o più rare del mare adriatico, IV. *Memorie R. Ist. veneto Sci.* **11**: 267–306.

ZINOVA, A. D. 1970. Novitates de algis marinis e sinu Czaunskensi (Mare Vostoczno Sibiriskoje dictum). *Nov. Sist. Nizsh. Rast.* **7**: 102–107.

INDEX TO GENERA

CPSIA information can be obtained at www.ICGtesting.com
Printed in the USA
BVOW07s0605030615

402948BV00008B/40/P

9 781907 807091